REVISE SALTERS-NUFFIELD AS/A LEVEL

Biology A

REVISION GUIDE

Series Consultant: Harry Smith

Author: Gary Skinner

A note from the publisher

In order to ensure that this resource offers high-quality support for the associated Pearson qualification, it has been through a review process by the awarding body. This process confirms that this resource fully covers the teaching and learning content of the specification or part of a specification at which it is aimed. It also confirms that it demonstrates an appropriate balance between the development of subject skills, knowledge and understanding, in addition to preparation for assessment.

Endorsement does not cover any guidance on assessment activities or processes (e.g. practice questions or advice on how to answer assessment questions), included in the resource nor does it prescribe any particular approach to the teaching or delivery of a related course.

While the publishers have made every attempt to ensure that advice on the qualification and its assessment is accurate, the official specification and associated assessment guidance materials are the only authoritative source of information and should always be referred to for definitive guidance.

Pearson examiners have not contributed to any sections in this resource relevant to examination papers for which they have responsibility.

Examiners will not use endorsed resources as a source of material for any assessment set by Pearson.

Endorsement of a resource does not mean that the resource is required to achieve this Pearson qualification, nor does it mean that it is the only suitable material available to support the qualification, and any resource lists produced by the awarding body shall include this and other appropriate resources.

For the full range of Pearson revision titles across KS2, KS3, GCSE, Functional Skills, AS/A Level and BTEC visit:
www.pearsonschools.co.uk/revise

Contents

- - - - - - - - - - - - - - - - - - - -

A small bit of small print
Edexcel publishes Sample Assessment Material and the Specification on its website. This is the official content and this book should be used in conjunction with it. The questions in *Now try this* have been written to help you practise every topic in the book. Remember: the real exam questions may not look like this.

Why is transport needed?

Mammals and other large organisms need a circulation system because they are too large for diffusion to be an effective way of supplying needs and removing waste. Specialised **organs**, like lungs and the heart, are required.

Surface area and volume

The **ratio** of the surface area of an organism to its **volume** affects how substances enter the organism, and how carbon dioxide leaves. You can use the abbreviation $\frac{SA}{V}$ to represent the surface area to volume ratio.

- **Small** organisms have a large $\frac{SA}{V}$. Substances and water can enter and leave by **diffusion** or **osmosis**.

- **Large** organisms have a smaller $\frac{SA}{V}$.

Maths skills — Small is big!

Small objects or mammals have a larger surface area to volume ratio, $\frac{SA}{V}$. Look at this cube.

Volume $= 1 \times 1 \times 1 = 1\,cm^3$

Surface area $= 6 \times 1 = 6\,cm^2$

So $\frac{SA}{V} = \frac{6}{1} = 6$

1 cm

Larger objects have a smaller $\frac{SA}{V}$. Look at this cube.

Volume $= 4 \times 4 \times 4 = 64\,cm^3$

Surface area $= 6 \times 16 = 96\,cm^2$

So $\frac{SA}{V} = \frac{96}{64} = 1.5$

4 cm

Maths skills Remember that at least 10% of your marks across all your exam papers will be maths-based.

Marvellous water

Water is a perfect solvent for transporting substances and carbon dioxide around the body because:

1 Water is a **polar molecule**.

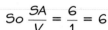

Polar molecules have **uneven** charge distribution. One end of this water molecule is slightly positive, δ^+, and the other slightly negative, δ^-. Water is said to be dipolar.

2 Water is **liquid** at **room temperature**.

 hydrogen bond

These three water molecules are joined to each other by 'sticky' hydrogen bonds, formed by the electrostatic attraction between the polar molecules.

3 Water easily **dissolves** molecules held together by ionic bonds.

Sodium and chloride ions in common salt (NaCl) are pulled apart by water molecules and then surrounded by them.

Water also dissolves polar molecules such as sugars and amino acids.

Worked example

The pictures show a pig and a flatworm.

Explain why the pig needs a circulatory system but the flatworm does not. **(2 marks)**

The flatworm has a large $\frac{SA}{V}$, whereas the pig has a smaller $\frac{SA}{V}$. The flatworm can meet its needs for food, oxygen and waste removal by diffusion. The pig cannot, so it needs a circulatory system to transport those substances around its body.

Now try this

Explain why water is described as a dipolar molecule. **(2 marks)**

If a question asks you to compare two things, give both sides of the story.

Blood vessels

The mammalian circulatory system is like a transport system. The blood vessels are like roads, the heart is like the motor (see page 3), and the blood is like the vehicles.

Structure and function in arteries, veins and capillaries

Vessel	Structure	Functional significance
artery	• relatively thick wall • smooth muscle • elastic fibres • lined with smooth layer of endothelial cells • narrow lumen	• withstands high blood pressure • alters diameter of lumen to vary blood flow • allow walls to stretch when blood is pumped into the artery and then recoil, smoothing blood flow • low friction surface to ease blood flow
capillaries	• very thin wall (just one cell thick)	• allows rapid exchange between blood and tissues
veins	• relatively thin wall • very little smooth muscle or elastic fibres • wide lumen • valves	• blood under low pressure • no pulse of blood so no stretching and recoiling • large volume acts as blood reservoir • stop backflow

Veins

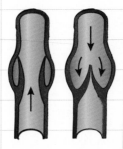

Veins have valves, which are not found in arteries or capillaries (except where arteries leave the heart). The valves prevent backflow, ensuring a one-way flow of blood toward the heart.

Cross-sections of blood vessels

artery   vein   capillary

☐ lumen   ■ outer coat   ■ muscle & elastic tissue   ■ endothelium

Arteries and veins are made of the same tissues but in different proportions. Capillaries only have endothelium.

Worked example

The photograph shows an artery and a vein.

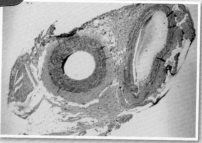

(a) Calculate how many times thicker the artery wall is than the vein wall at the points shown by the lines.

Vein wall measures 1 mm.

Artery wall measures 4 mm, so it is 4 times thicker.

(b) The magnification is ×100. Calculate the actual thickness of the muscle and elastic tissue region of the artery.

Actual measured thickness of artery muscle and elastic tissue is 4 mm.

so, $100 = \dfrac{4}{\text{actual thickness}}$

actual thickness $= \dfrac{4}{100}$

$= 0.04\,\text{mm}$

📟 **Maths skills** The following equation shows the relationship between magnification and image sizes:

$\text{magnification} = \dfrac{\text{image size}}{\text{actual size}}$

Remember that at least 10% of your marks across all your exam papers will be maths-based.

Now try this

Describe **two** differences between the structure of a capillary and the structure of a vein.   **(2 marks)**

The heart

The heart consists of four chambers (pumps). The right and left sides are completely separate, but there are valves between the top chambers (atria) and bottom chambers (ventricles).

A simple model of the four-chambered heart

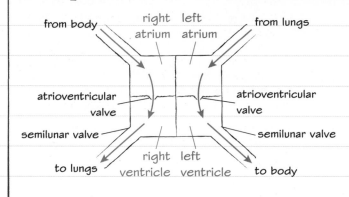

from body   right   left   from lungs
            atrium   atrium

atrioventricular   atrioventricular
valve   valve

semilunar valve   semilunar valve

to lungs   right   left   to body
            ventricle ventricle

You need to know the structure of the heart and understand how the structure relates to how it functions. It might help you to think of the heart as a box with four chambers and four pipes: two going in and two coming out. In addition, the top two chambers are connected to the bottom two by valves.

A simple model of the four-chambered heart. Deoxygenated blood (blue) enters the right atrium from the body. It is then pushed, by muscles in the right ventricle, to the lungs where it is oxygenated. The oxygenated blood (red) returns from the lungs to the left atrium, is pushed into the left ventricle and then out along the main blood vessel of the body, the aorta.

The anatomical structure of the heart

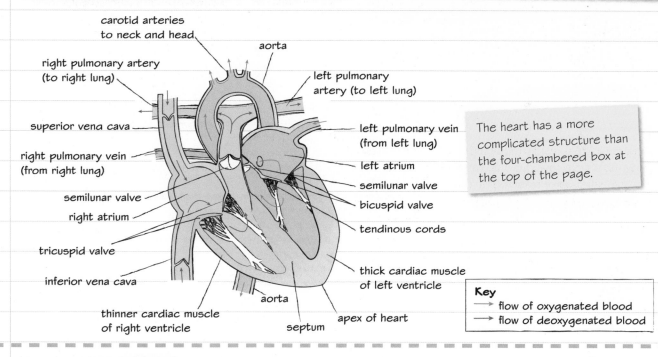

carotid arteries
to neck and head

aorta

right pulmonary artery
(to right lung)

left pulmonary
artery (to left lung)

superior vena cava

left pulmonary vein
(from left lung)

right pulmonary vein
(from right lung)

left atrium

semilunar valve

semilunar valve

bicuspid valve

right atrium

tendinous cords

tricuspid valve

inferior vena cava

thick cardiac muscle
of left ventricle

thinner cardiac muscle
of right ventricle

aorta

septum

apex of heart

The heart has a more complicated structure than the four-chambered box at the top of the page.

Key
→ flow of oxygenated blood
→ flow of deoxygenated blood

Worked example

Explain why the mammalian heart has two sides.   **(2 marks)**

You need to make sure you relate structure to function.

Two sides allow oxygenated and deoxygenated blood to stay separate, which allows as much oxygen as possible to be carried to the cells.

Having two sides allows a different amount of muscle on each side, so that there can be a higher pressure on one side (pumping to the body) compared to the other (pumping blood to the lungs).

Now try this

Explain the difference in thickness of the wall of the right atrium and the wall of the right ventricle.   **(3 marks)**

The cardiac cycle

The heart goes through a cyclical process about 60 to 70 times a minute, receiving oxygenated blood from the lungs and pumping it around the body.

The cardiac cycle in the left side of the heart

Blood drains into left atrium from lungs along the pulmonary vein.

↓

Raising of the blood pressure in the left atrium forces the left atrioventricular valve open.

Contraction of the left atrial muscle (left atrial systole) forces more blood through the valve.

↓

As soon as left atrial **systole** (muscle contraction) is over, the left ventricular muscles start to contract. This is called left ventricular systole.

↓

This forces the left atrioventricular valve to close and opens the valve in the mouth of the aorta (semilunar valve). Blood then leaves the left ventricle along the aorta.

The same steps are repeated on the right side at the same time.

🧪 Practical skills — Heart dissection

You can relate structure to function in the heart when you do a dissection.

- The heart has its own blood vessels called the coronary arteries and veins, which can be seen branching over its surface.
- Water run into the pulmonary vein will emerge from the aorta, showing the heart is separated into two halves internally.
- When cut, the left and right ventricles are very different in thickness. This is related to their different roles (see page 3 for a reminder).
- Inside you can see the atrioventricular valves that stop blood flow from ventricles to atria and semilunar valves from arteries back into the heart.
- Attached to the atrioventricular valves you can see tendons (the heart strings) which stop the valves turning inside out.

ECG of heart beat

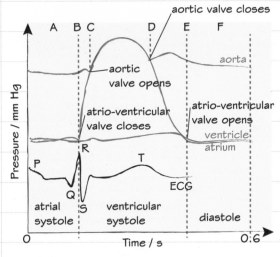

B: ventricle starts to contract, ventricular pressure greater than atrial pressure so atrioventricular valve closes

C: the pressure in the ventricles is now greater than that in the aorta so the semilunar valve (aortic value) opens

D: ventricular pressure in now lower than that in the aorta so the semilunar valve closes

E: the ventricular pressure is now lower than the atrial pressure so the atrioventricular valve opens and blood flows from atrium to ventricle

F: blood still entering atria from pulmonary vein and moving on into the ventricle

In the ECG, the P wave is when electrical excitation spreads over the atria and the QRS and T waves are when electrical excitation is spreading over the ventricles

Ventricles are contracting in systole. The valves between the ventricles and the atria will be closed. The SAN has sent out its signal. Muscle contractions makes blood pressure high.

Worked example

Which column is correct for ventricular systole? Tick the box.   **(1 mark)**

Three features	☐	☐	☐	☑
atrioventricular valves	closed	open	open	closed
SAN pacemaker	active	not active	active	not active
ventricular blood pressure	lower than atria	lower than atria	higher than atria	higher than atria

Now try this

Describe the roles of the atrioventricular (bicuspid and tricuspid) valves during the cardiac cycle.   **(4 marks)**

Clots and atherosclerosis

Blood clots form when there is damage to tissue. If there is damage to an artery and a clot forms this can cause many problems.

Atherosclerosis

Atherosclerosis is the disease process that leads to coronary disease and strokes (cardiovascular diseases). Fatty deposits (**atheroma**) can either block an artery directly or increase its chance of being blocked by a blood clot (**thrombosis**).

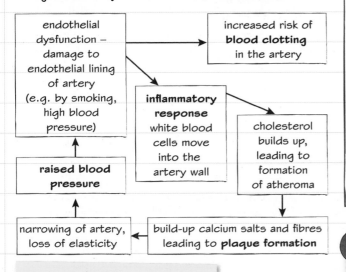

Development of atherosclerosis

Atheroma

Health effects of atheroma:

- **increased blood pressure**, which can damage the kidneys and retina and cause strokes
- **aneurysm**, where the increased pressure of blood caused by the blockage can lead to the bursting of an artery and internal bleeding
- **angina**, a chest pain often felt during exercise, caused by reduced blood flow to heart due to narrowing of coronary arteries
- **heart attack**, when a coronary artery becomes totally blocked, usually by a clot, and part of the heart is starved of oxygen and dies
- **stroke**, an interruption to the blood supply of the brain which can cause paralysis or even death.

The clotting cascade

Because it is crucial that a blood clot doesn't form in the wrong place (such as in blood vessels) or at the wrong time, a number of factors have to be present before it will happen.

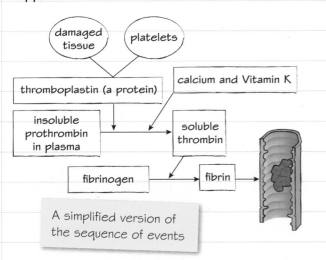

A simplified version of the sequence of events

Worked example

Frequent use of cocaine leads to an increase in the level of von Willebrand factor, a protein, in the blood. This protein naturally assists in blood clotting. The diagram shows how.

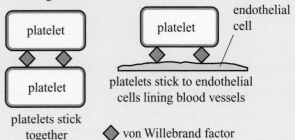

platelets stick together

platelets stick to endothelial cells lining blood vessels

◆ von Willebrand factor

Explain why cocaine use may increase the chance of a blood clot forming.   **(4 marks)**

The von Willebrand factor causes platelets to stick to each other and the endothelium. They release thromboplastin and trigger blood clotting.

There will be more fibrinogen and this will result in the clot growing larger.

The 'sticky' platelets cause blood flow to decrease.

If the blood is flowing slower then there is a further increase in the chance of blood clotting.

Now try this

Describe how atherosclerosis develops.   **(4 marks)**

Risk

Risk is the probability of the occurrence of an unwanted event or outcome.

Perceptions of risk

People often overestimate risk of something if it is:

- involuntary (being a passenger in an airplane as opposed to driving a car)
- dreaded (nuclear accident, natural disaster)
- not natural (nuclear power)
- unfair
- unfamiliar
- very small

Poor diet, lack of exercise and smoking increase the risk of cardiovascular disease (CVD) but because of problems with the risk perception mentioned above, people underestimate the risk of CVD and do not change their lifestyle.

Risk and heart disease

In this case risk is poorly judged because of:

- own experience, which will carry more weight than statistics
- inability to assess risks well
- peer pressure, for example alcohol consumption and smoking among young people
- the idea that if something is destined to happen then it will and there is not much one can do about it (karma)
- the remoteness of the likely consequence.

 Maths skills **Factors that increase the risk of CVD**

many correlations between dietary habits and levels of CVD, e.g. lipoprotein and salt levels; some might be causal, particularly for blood cholesterol levels

high blood pressure — very important – should not be above 140 mm Hg systollic and 90 mm Hg diastolic

genetics — inherit tendency to: high blood pressure; poor cholesterol metabolism; arteries that are more easily damaged; relative HDL : LDL levels in blood

diet

risk factors for CVD

correlation and causation shown because chemicals in smoke physically damage artery linings and also cause them to constrict — *smoking*

inactivity — regular vigorous exercise reduces the risk of CVD by reducing blood pressure and raising HDL (good cholesterol) levels

oestrogen gives women some protection from CVD before the menopause — *gender*

age — elasticity and width of arteries decrease with age

If we assume a population of 64 million and take the figure of 100 000 dying from heart disease each year, the probability of so many dying is $\dfrac{100\,000}{64\,000\,000} = 0.0015$ or 0.15%.

This is about 1 in 660.

Worked example

Analyse the data in the graph to describe the trends shown. **(4 marks)**

The trend is a decrease in smoking levels for both men and women. Women always smoke less than men.

The fall in the levels of smoking was faster in men between 1974 and 1986.

The overall fall in men smoking was 31% and in women 24%.

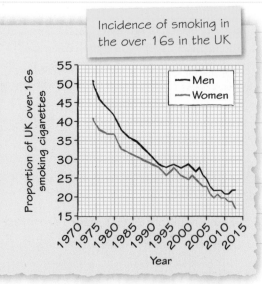

Incidence of smoking in the over 16s in the UK

Now try this

Explain the effects of high blood cholesterol, high blood pressure and smoking on the onset of CVD. **(5 marks)**

Correlation and causation

Two variables can be **correlated** when a change in one is accompanied by a change in the other, but correlation does not always mean **causation**. When the change in one variable is the cause of the change in the other, they are **causally linked**. Epidemiologists look for correlations and then further work needs to be done to see if a mechanism to explain why one may cause the other can be found.

Correlation studies

These studies allow you to determine if there is a **correlation** (or not) between a risk factor (such as blood cholesterol) and health.

Graphs like this can be produced:

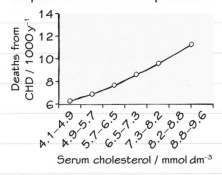

Relationship between blood cholesterol levels and death from CHD (coronary heart disease, a type of CVD)

Such graphs show a correlation between high blood cholesterol and CHD. However, it does not give evidence that high blood cholesterol **causes** CHD.

Causation

Studies have shown a strong correlation between high intake of saturated fats and high blood cholesterol. Cholesterol helps form plaques; scientists believe that this is the **causal link** (for more about this study see page 13).

Another famous case of a correlation between a health problem and a possible risk factor is that of smoking and lung cancer. The causal link is to do with mutations caused by chemicals in smoke.

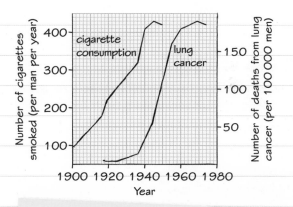

The number of cigarettes smoked per year and deaths from lung cancer, 1900–1980

Worked example

(a) Describe how the graph above right shows evidence for a correlation between smoking and deaths from lung cancer. **(2 marks)**

The change in the number of deaths from lung cancer is similar to the change in the number of cigarettes smoked.

Also, the change in number of deaths is approximately the same number of years after the changes in cigarette smoking.

(b) Does it show evidence that smoking causes lung cancer? **(1 mark)**

No

Now try this

Explain what needs to be done to show that smoking causes lung cancer. **(3 marks)**

Studying the risks to health

Epidemiologists are scientists who carry out research on patterns of diseases or health risks in populations to try to determine the risk factors for health.

There are two main kinds of study

Cohort studies:

- follow a large number of people over an extended period
- subjects are monitored to see if they develop the condition
- cohort then divided into groups, those with and those without the condition
- subjects interviewed to assess their risk factors
- correlation between risk factors and development of the condition is looked for.

Case–control studies:

- a group with the condition (cases) is compared with a group without it (control)
- past histories of the two groups are investigated
- the study will only have validity if the two groups are matched for other factors such as age and gender.

What makes a good study?

In all epidemiological studies various measures have to be taken to ensure the collection of adequate results.

✓ Variables should be controlled when selecting cohorts or control groups to ensure validity and reliability. This is difficult because human beings are so variable.

✓ Measurement techniques or the questions on a questionnaire must be standardised.

✓ Sample size is very important. For many diseases, only a low percentage of the population has the condition so an apparently large sample size might only contain a small number of individuals with the condition.

✓ The studied sample should be representative of the whole population to avoid bias.

Worked example

Researchers collected data to study the relationship between the time spent watching television and coronary heart disease (CHD). None of the subjects had previously had a stroke or heart attack.

(a) Explain why people who had not had strokes or heart attacks were selected for this study. **(2 marks)**

To standardise the health of the sample group. It also increases confidence that any CHD developed during the period of the investigation.

The participants were asked questions about smoking, family history of CVD, alcohol intake, sleep duration, total energy intake, physical activity, medication and time spent watching television. In addition, data on height and body mass, blood pressure, HDL and LDL cholesterol, waist circumference and plasma triglycerides were collected by health professionals.

(b) Explain why the data collected by the health professionals might be considered to be more valid than the information in the questionnaire. **(4 marks)**

The data collected by health professionals is objective and quantitative. The health professionals are trained and are likely to be less biased.

With the questionnaire, participants may over- or under- estimate and have to rely on remembering events and medical history accurately.

Now try this

The graph shows the percentage of the male population (M) and the female population (F) who are either overweight or obese in five different countries.

(a) Calculate the ratio of overweight males to overweight females in the USA. **(1 marks)**

(b) Comment on any correlation between gender and being overweight or obese that the data show. **(2 marks)**

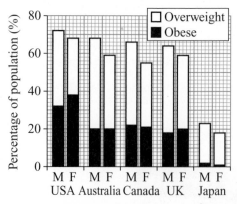

The data are percentages not numbers and we are not told sample sizes.

Energy budgets

Energy budget is the balance between the amount of energy a person requires compared to the energy they take in through their diet.

Energy budgets

The energy you need in a day depends on your **basal metabolic rate** (BMR) and level of activity. BMR is usually quoted as the basic energy requirement for a day. It varies with gender, age and body mass. If the energy transferred from the food you eat is the same as your BMR, your weight will stay the same. An energy imbalance puts you at risk of weight loss or gain and obesity. This balance is an energy budget.

Fred's daily energy intake

1800 kJ breakfast   4000 kJ supper

2500 kJ lunch   2000 kJ drinks and snacks

Total: 10 300 kJ day^{-1}

 Maths skills ## Calculating energy budgets

Fred, a carpenter, weighs 70 kg. The BMR for a man of Fred's age is 4 kJ kg^{-1} body mass h^{-1}. So Fred's basic energy requirement is $4 \times 70 = 280$ kJ h^{-1}.

His BMR per day is: $280 \times 24 = 6720$ kJ day^{-1}.

Daily activity for an average carpenter uses 5000 kJ.

Fred's total energy need per day is therefore:

$6720 + 5000 = 11\,720$ kJ

Fred's daily energy intake (see post it) is 10 300 kJ,

so his daily energy balance is:

$10\,300 - 11\,720 = -1420$ kJ.

He is using more energy than his intake, so he will lose weight.

Bob, another male carpenter of the same age and weight as Fred, consumes 13 700 kJ day^{-1}.

$13\,700 - 11\,720 = 1980$ kJ

Bob will put on weight and could become obese.

 Maths skills ## Body mass index

Around 50% of the UK adult population is overweight, 20% are obese. Body Mass Index (BMI) is an indicator used to define these conditions.

$$BMI = \frac{\text{body mass in kg}}{(\text{height in m})^2}$$

A person who is 1.75 m tall and weighs 70 kg, would have BMI:

$$BMI = \frac{70}{1.75^2} = \frac{70}{3.06} = 22.9$$

The value can be looked up in a table.

BMI	Status
<20	underweight
20–25	correct weight
25.1–30	overweight
>30	obese

Waist : hip ratio

This is another indicator of obesity. A waist : hip ratio should not be greater than 0.9 in men and 0.85 in women.

Maths skills Ratios do not have units, but the values should both be measured in the same units.

Worked example

The graph shows the relative mortality against BMI for men and women.

Analyse these data to describe the effect of BMI on relative mortality in men and women. **(4 marks)**

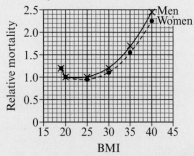

Relative mortality decreases as BMI increases from 19 to 21 in both men and women.

There is little change in relative mortality within the range 20 to 25 in both sexes.

As BMI increases from above 25 the risk increases in both men and women, but above 25 the risk for men is greater than that for women.

Now try this **Maths skills**

Given that a 100 kg man has a BMI of 25, calculate his height. **(2 marks)**

Monosaccharides and disaccharides

'Carbo' refers to carbon (C) and 'hydrate' to water (H_2O). Carbohydrates are substances made of C and H_2O with the general formula $C_x(H_2O)_n$. and they are the main source of energy in the diet.

Types of saccharides

Saccharides (shown as ⬡) are sugar units that make up carbohydrates.

- a monosaccharide has one unit
- a disaccharide has two units ⬡⬡
- polysaccharides have many units (see page 11).

You need to know the difference between these three kinds of saccharide.

> **Be careful!** Do not use the word sugar without qualification in a science context. In everyday use, 'sugar' means sucrose or table sugar, but in science there are hundreds of sugars.

Monosaccharides

Glucose ($C_6H_{12}O_6$) is a hexose sugar because it has 6 (= hex) carbon atoms.

The molecule is nearly always in the form of a ring.

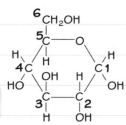

alpha α-glucose

Glucose structure and functions

- Glucose is the source of the most readily available energy from respiration in living things.
- Its solubility in water helps it fulfil this role as it can easily be carried to where it is needed.
- It is also a relatively small molecule which is important for its movement into cells.

Making and splitting disaccharides

Disaccharides are made when two monosaccharides join by losing water (**condensation reaction**). When they join at C atoms 1 and 4, they form a 1,4 glycosidic bond.

Disaccharides are split by adding water (**hydrolysis reaction**). Water is needed and the glycosidic bond is broken.

A 1,4 glycosidic bond (red) in maltose

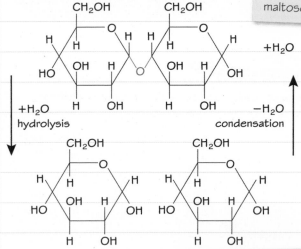

+H_2O
hydrolysis

−H_2O
condensation

Condensation and hydrolysis of two α-glucose molecules

Examples of disaccharides

- **Maltose** is two α-glucose molecules.
- **Sucrose** is an α-glucose molecule and a fructose molecule.

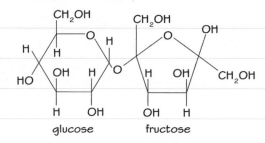

glucose   fructose

- **Lactose** (milk sugar) is glucose and galactose joined with a β 1,4 bond because the glucose is the β form.

glucose   galactose

(see page 11).

Worked example

A compound with chemical formula $C_{12}H_{22}O_{11}$ could be

A glucose, **B** fructose, **C** ribose, **D** maltose   (**1 mark**)

D maltose

> **A** and **B** are hexoses ($C_6H_{12}O_6$) and **C** is pentose ($C_5H_{10}O_5$). Which other molecules on this page would have the formula $C_{12}H_{22}O_{11}$?

Now try this

Why does sucrose (two hexose sugars ($C_6H_{12}O_6$) joined together) not have chemical formula $C_{12}H_{24}O_{12}$?   (**2 marks**)

Carbohydrates – polysaccharides

When many monosaccharides join together they form a **polysaccharide**. The monosaccharides are **monomers** and the polysaccharide is the **polymer** that forms when they join.

Polysaccharide structure

Polysaccharides have many sugar units. They can be **unbranched** as in amylose (a component of starch), or they can be **branched**, as in amylopectin (another starch component) and glycogen (animal starch).

Unbranched polysaccharide

Branched polysaccharide

Joining of monosaccharides to make polysaccharides

- α-glucose monomers join with α 1,4 glycosidic bonds to make amylose
- α-glucose monomers join with α 1,4 and α 1,6 glycosidic bonds to make amylopectin and glycogen
- glycogen has more α 1,6 bonds than amylopectin, it is more branched
- starch is a mixture of amylopectin and amylose.

When starch is broken down in digestion, the enzymes catalyse a hydrolysis reaction and monosaccharides are produced again.

The structure of α-glucose polymers

In **amylose** the straight chain coils up due to hydrogen bonding.

In **amylopectin**, the branches (1,6 glycosidic bonds) do not allow so much coiling.

Glycogen is like amylopectin but with more frequent branches.

Structure and function in polysaccharides

Starch:

- the coiled shape makes it compact, so it can store lots of glucose and therefore energy in a small space
- it is insoluble so does not exert a problematic osmotic effect
- amylopectin is more easily broken down than amylose due to more branches; this creates more terminal ends where breakdown occurs, and gives a mixture of rapid and slower glucose and energy release.

Glycogen:

- has even more branches so can release glucose even faster, which is needed in animals.

Worked example

Distinguish between the structures of amylose and amylopectin.   **(2 marks)**

Amylose is straight-chained whereas amylopectin is branched.

Amylose has 1,4 glycosidic bonds whereas amylopectin has both 1,4 and 1,6 glycosidic bonds.

If asked to compare two things, make sure you mention the situation in **both**.

Now try this

Put a tick in the box where you think the molecule has the feature shown.   **(3 marks)**

Structural feature	Sucrose	Amylopectin
glycosidic bonds present		
side chains present		
contains more than three glucose monomers		

Lipids

Lipids (fats and oils) are all hydrophobic ('water-hating') molecules. They are made up of fatty acids and glycerol but have a variety of structures. Lipids are used to store energy, and act as waterproofing and insulating agents. They are a very important part of all membranes around and inside cells.

Triglycerides

Triglycerides are made of three (tri) fatty acids and a glycerol joined by **ester bond**s. These are formed by a **condensation reaction** in which water is lost.

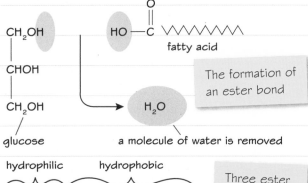

fatty acid

The formation of an ester bond

glucose   a molecule of water is removed

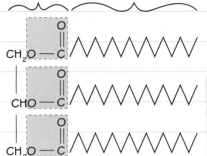

Three ester bonds in a triglyceride

the tryglyceride has a hydrophobic tail and a hydrophilic head

Fatty acid facts

Lipids vary because fatty acids vary:
- fatty acids are of different lengths
- in mixed triglycerides the three fatty acids are different from each other
- fatty acids may be **saturated** or **unsaturated** (with hydrogen).

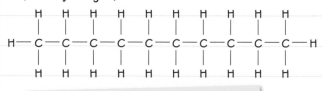

A saturated fatty acid with no double bonds between carbon atoms

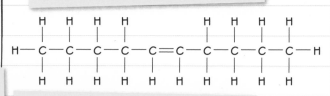

An unsaturated fatty acid with double bonds between some carbon atoms

Properties of saturated and unsaturated fatty acids

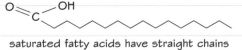

saturated fatty acids have straight chains

Palmitic acid (saturated)

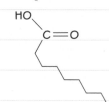

Oleic acid (unsaturated)

unsaturated fatty acids have a kink at each double bond

The kink in an unsaturated fatty acid makes the membrane more fluid than one with saturated lipids.

Draw a mixed triglyceride made up of the following components.

(3 marks)

Lipases are enzymes that are involved in the breakdown of triglycerides.

(a) Name the bond broken by lipases. **(1 mark)**

(b) Explain how the breakdown of triglycerides would affect the pH of a reaction mixture.

(2 marks)

Good cholesterol, bad cholesterol

Cholesterol is an essential component of cell membranes where it affects their fluidity.

Cholesterol

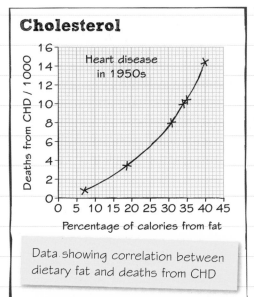

Heart disease in 1950s

Data showing correlation between dietary fat and deaths from CHD

However, it is now known that there are two types of cholesterol, 'good' and 'bad'.

Good and bad cholesterol

Cholesterol is a water-insoluble lipid which has to be carried by proteins in complexes called lipoproteins.
There are two kinds of lipoprotein.

Low density lipoprotein (LDL, 'bad' cholesterol)	High density lipoprotein (HDL, 'good' cholesterol)
formed from saturated fats, protein, cholesterol	formed from unsaturated fats, protein, cholesterol
bind to cell surface receptors, which can become saturated, leaving the LDLs in the blood	transport cholesterol from body tissues to liver where it is broken down.
associated with formation of atherosclerosis	reduces blood cholesterol levels, discourages atherosclerosios
should be maintained at low level	should be maintained at high level

The LDL/HDL ratio is important in determining risk of cardiovascular problems

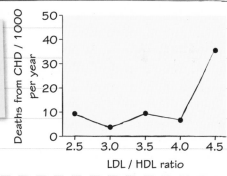

The causal link

The fact that cholesterol helps form plaques could be the causal link between the two variables. Rats fed on a high cholesterol diet formed more plaques in the endothelium of arteries than the control group. As long as it is clear that all other possible variables are controlled and are the same in both groups, this then leads further to saying that there is a causal relationship. A mechanism has now been found which explains how lipoproteins cause plaques.

The Framingham Study

Data about the cholesterol and CVD link.

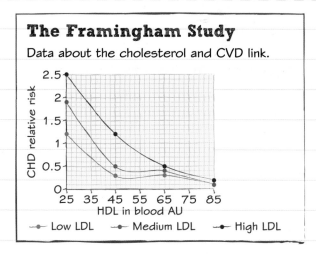

—•— Low LDL   —•— Medium LDL   —•— High LDL

Use the line graph to explain the interaction between cholesterol and CVD. **(3 marks)**

Decreasing HDL always leads to an increase in CVD. Whereas decreasing LDL always leads to a decrease in CVD. The very lowest risk of CVD is found when LDL levels are low and HDL levels are high.

Now try this

(a) Explain why you would expect CHD rates in an area with higher rates of alcohol consumption and regular strokes to be higher than those in an area with low alcohol consumption rates and fewer strokes. **(2 marks)**

(b) Suggest why this may not be the case. **(2 marks)**

Reducing the risk of CVD

People can use some of the scientific evidence to help to reduce their risk of CVD by changing their lifestyle.

Reducing the risk of CVD

The most important actions to be taken are:

- stop smoking
- maintain resting blood pressure below 140/90 mm Hg
- maintain low blood cholesterol level
- maintain a normal BMI / low waist-to-hip ratio (see page 9 for a reminder of BMI and W : H)
- take regular physical exercise
- moderate or no use of alcohol
- reduce stress.

Dietary strategies

- reduce saturated fats (reduces total cholesterol, but LDL : HDL more)
- more polyunsaturated fats (reduces LDL : HDL)
- reduce salt (lower fluid levels, in turn reduces blood pressure)
- more fruit and vegetables (antioxidants and non-starch carbohydrates, which lower blood cholesterol)
- more oily fish (linked to slight reduction in blood pressure and risk of blood clotting)
- more non-starch polysaccharides.

Be careful! Relate the strategies suggested above to reduce CVD, using scientific knowledge and evidence.

Vitamin C

Vitamin C is an antioxidant common in many foods such as citrus fruits; antioxidants are important in neutralising free radicals which can damage cells. It is destroyed by heat treatment.

The Vitamin C content of food and drink is easy to find out with a simple practical procedure based on the following:

- antioxidants are reducing agents since they lose electrons
- many substances will change colour when they are reduced; these substances are called **redox dyes** (reduction–oxidation dyes)
- DCPIP is a blue dye in its non-reduced form; it goes colourless when in the presence of Vitamin C as it gains electrons.

 **Practical skills** This is **core practical 2**. Notice that the more juice that is required to decolourise the DCPIP the less Vitamin C it contains.

Worked example

Assume it took 0.6 cm³ of a 1% Vitamin C solution to decolourise some DCPIP.

(a) Calculate the mass of Vitamin C in 0.6 cm³ of a 1% solution. **(2 marks)**

In a 1% Vitamin C solution, there is 1 g of Vitamin C in 100 cm³ of liquid, so there are 10 mg of Vitamin C in 1 cm³. In 0.6 cm³ of the solution, there is 6 mg of Vitamin C.

(b) Use these results to calculate the mass of Vitamin C in the grapefruit juice. **(2 marks)**

Juice tested	Average volume of juice required to decolourise DCPIP / cm³
grapefruit juice	1.61
pineapple juice	11.56
orange juice	2.12
orange drink	1.45
fresh lemon juice	1.73
bottled lemon juice	24.00

6 mg Vitamin C decolourised 1 cm³ of DCPIP.

1.61 cm³ of the grapefruit juice decolourised 1 cm³ of DCPIP so 1.61 cm³ of grapefruit juice contains 6 mg Vitamin C.

Therefore 1 cm³ of grapefruit juice contains:
$\frac{6}{1.61} = 3.7$ mg of Vitamin C

Now try this

Calculate the Vitamin C content per cm³ of pineapple juice from the table. **(2 marks)**

Medical treatments for CVD

Changes in lifestyle and diet may not be enough to prevent CVD in some people; they might need drugs to control their blood pressure and or lower blood cholesterol.

Drugs to reduce the risk of CVD

Drug treatment	Mode of action	Risk/side effect
diuretics (antihypertensive)	increases volume of urine; lowers blood volume and blood pressure	very occasional dizziness, nausea, muscle cramps
calcium channel blockers (antihypertensive)	disrupts calcium ion movement through calcium channels in the cell membrane, reducing muscle contraction, so increases diameter of arteries, reduces force of heart beat and frequency, lowering blood pressure	headaches, dizziness, swollen ankles, constipation and flushing in the face
ACE (angiotensin converting enzyme) inhibitors (antihypertensive)	blocks the production of ACE, reducing arterial constriction and lowering blood pressure	cough, dizziness, abnormal heart rhythm, impaired kidney function
statins	inhibit an enzyme in the liver that produces LDL cholesterol	tiredness, disturbed sleep, nausea, diarrhoea, headache, muscle weakness; also people may depend wholly on statins and neglect to eat a healthy diet
anticoagulants, e.g. warfarin	reduce risk of clot formation	risk of uncontrolled bleeding; dosage control is essential
platelet inhibitory drugs, e.g. aspirin, clopidogrel	make platelets less sticky	aspirin irritates the stomach lining and can cause stomach bleeding; clopidogrel with aspirin increases this risk

Worked example

Two groups of patients were treated with a different type of cholesterol-reducing drug, Drug A or Drug B. The graphs below show the percentage changes of total cholesterol (TC), low density lipoproteins (LDL) and high density lipoproteins (HDL) in the blood of these patients, after treatment.

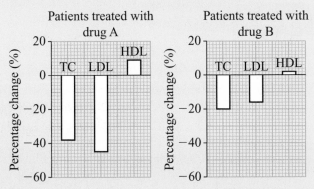

The enzyme HMG-CoA reductase catalyses the synthesis of cholesterol. When this enzyme is active, there are fewer LDL receptors on liver cells. These receptors are needed to remove LDL from the blood.

(a) Compare and contrast the effects of Drug A and Drug B on the percentage changes in total cholesterol (TC), LDL and HDL in the blood of these two groups of patients. **(2 marks)**

Both drugs have an effect on all blood parameters. However A has a greater effect than B on them all.

(b) Statins are a group of cholesterol-reducing drugs that act by inhibiting this enzyme. Explain which of the two drugs, Drug A or Drug B, is more likely to be a statin. Give reasons for your answer. **(4 marks)**

Drug A because the total cholesterol and LDL levels are lower and statins inhibit cholesterol synthesis.

In addition statins result in more LDL receptors on liver cells so more LDL will be cleared from the blood.

Now try this

Describe the risks of using statins to treat CVD. **(2 marks)**

15

Daphnia heart rate

 Practical skills Some dietary components, such as caffeine, may have an effect on the heart, but this is not easy to study in humans for ethical reasons. **Core practical 1** uses a water flea (*Daphnia*).

Ethical issues

Daphnia	Human
simple nervous system	complex nervous system
no need for dissection, as transparent and can see heart beating	needs dissection
abundant in nature	cannot kill for experiment
bred for fish food	
no loss of genetic variation	
cannot give consent	can give consent

Why use *Daphnia* for this practical?

☑ abundant

☑ easily obtained

☑ transparent, so heart can be seen

☑ simple nervous system, so ethically less of an issue than a mammal.

Some limitations

Although simple animal models such as *Daphnia* are useful to get ideas about how human body systems work, it must be remembered they are not human. Conclusions drawn from such experiments should be considered carefully when extended to humans.

Ensuring a successful experiment

Step in experiment	Method
immobilise the *Daphnia*	use strands of cotton wool in a small dish of the experimental solution to trap the *Daphnia*
control other variables, such as water temperature, *Daphnia* size, etc.	difficult to maintain a constant temperature but it should be monitored with a thermometer in the water; *Daphnia* of similar size, etc. should be used for all experiments
accurate measurement of heart rate	dots are put on a piece of paper (in an S shape to avoid putting one dot on top of another) or repeatedly press a button on a calculator
repeatability	ensure that variables other than caffeine concentration are controlled

Variables and safety

The variables to be considered are:

☑ temperature

☑ age, size and sex of *Daphnia*

☑ aspects of pretreatment such as type of water, length of time out of natural habitat.

The safety measures to be taken are:

☒ do not mix water and electricity

☒ don't forget to wash your hands after handling the *Daphnia*.

Worked example

In an experiment to look at the effect of ethanol on heart rate, *Daphnia* were placed in solutions of different concentrations of ethanol kept at 25 °C. Their heartbeats were observed using a microscope and the heart rates were recorded.

Explain why the temperature was maintained at 25 °C in this investigation. **(3 marks)**

This was done so that the only variable which changed was the ethanol concentration.

This avoids heart rate changing due to a change in temperature.

25 °C was chosen to ensure a good level of *Daphnia* activity without leading to enzyme denaturation.

Now try this

Explain ways in which the accuracy of this experiment might be improved. **(4 marks)**

Exam skills

This exam-style question uses knowledge and skills you have already revised. Have a look at pages 12, 13 and 14 for a reminder.

Worked example

(a) The diagrams below show part of the structure of the surface of high density lipoprotein (HDL) and low density lipoprotein (LDL).

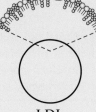

 phospholipid

⬭ protein

cholesterol

HDL   LDL

Using the information in the diagram, compare and contrast the structure of HDL and LDL. **(4 marks)**

> Compare and contrast requires you to look for the similarities and differences of two (or more) things. The answer must relate to both (or all) things mentioned in the question and must include at least one similarity and one difference.

Both contain phospholipid, protein and cholesterol.

HDL is smaller with more protein and less cholesterol.

(b) A study was carried out to determine the relative risk of developing cirrhosis of the liver in relation to the mass of alcohol consumed each day by men and women. The graph below shows the results. Compare and contrast the effects of alcohol consumption on the risk of cirrhosis in men and women. **(3 marks)**

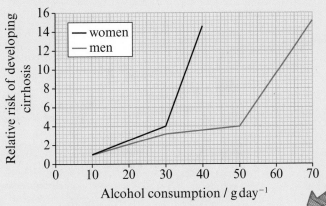

> If you are asked to do something with numerical information, such as in this graph, make sure you quote and/or manipulate the numbers. It would seem easy here to simply say that women have a greater risk of cirrhosis than men, but this is not true at the lower consumption levels. The other point in the answer is that it supports the simple conclusion that women have a greater risk than men, using data from the graph.

It increases the risk in both men and women.

Women have a greater risk than men, although there is little difference between men and women at lower alcohol consumption levels (10–30 g day^{-1}). The steeper increase in risk begins at a lower alcohol consumption in women (30 g day^{-1}), compared to men (50 g day^{-1}).

(c) The table below shows the mean concentration of triglyceride, LDL and HDL in the blood of people with and without cirrhosis of the liver.

Type of lipid	Mean concentration of lipid in blood / mg dm^{-3}	
	People with cirrhosis	People without cirrhosis
triglyceride	109	92
LDL cholesterol	131	103
HDL cholesterol	48	49

Analyse the information in the table to explain why people suffering from cirrhosis may also be more at risk of developing cardiovascular disease (CVD). **(4 marks)**

> In this answer, 'analysing the information' has been interpreted as actually doing a calculation, which can then be used to explain the different risks.

People with cirrhosis have higher LDL (cholesterol) levels. Possibly more importantly, their HDL:LDL cholesterol ratio is 0.37 compared to 0.48 for those without cirrhosis.

A lower HDL:LDL may mean that LDL overloads membrane receptors, resulting in atherosclerosis.

Gas exchange

Living things rely on **diffusion** to move gases in and out of their bodies (gas exchange). Their anatomy and physiology reflect this reliance.

Features of gas exchange surfaces

Gas exchange surfaces in living things have features that maximise the rate of diffusion:

- a large surface area to volume ratio (see page 1)
- a thin barrier to movement (thin surface)
- a steep concentration gradient (difference in concentration) maintained.

Maths skills — Fick's law

These features of gas exchange surfaces are combined in **Fick's law**, expressed in this equation:

$$\text{rate of diffusion} \propto \frac{\text{surface area} \times \text{difference in concentration}}{\text{thickness of gas exchange surface}}$$

Notice the 'proportional to' sign ($\propto$). To calculate the rate, you would need a **diffusion coefficient**, which is a constant under standard conditions. The coefficient depends on what is diffusing, temperature and pressure.

The human lung

Fick's law means that diffusion is faster when surfaces have a large area and are thin. In addition, the steeper the concentration gradient, the faster diffusion will be.

Lungs are adapted for rapid gas exchange. They have about 600 million air sacs (**alveoli**) giving an area of about $100\,m^2$. The alveoli have a wall which is made up of a single thickness of flat cells, so it is very thin. They are filled with fresh air (high O_2, low CO_2) about 15 times a minute.

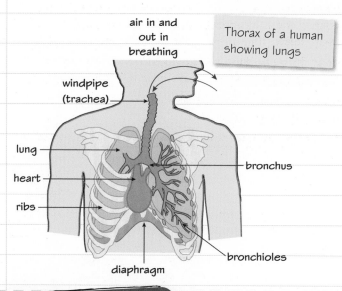

Thorax of a human showing lungs

air in and out in breathing

windpipe (trachea)

lung

heart

ribs

diaphragm

bronchus

bronchioles

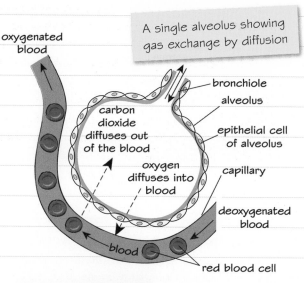

A single alveolus showing gas exchange by diffusion

oxygenated blood

carbon dioxide diffuses out of the blood

oxygen diffuses into blood

bronchiole

alveolus

epithelial cell of alveolus

capillary

deoxygenated blood

blood

red blood cell

Worked example

The photograph shows a single-celled organism called *Amoeba*.

magnification ×800

You are told that it is magnified ×800, so this tells you *Amoeba* is very small.

Explain how *Amoeba* carries out gas exchange. **(4 marks)**

Gas exchange occurs by diffusion through the cell membrane which is thin.

Diffusion is sufficient because *Amoeba* has a large surface area to volume ratio.

Now try this

The diagram shows a small part of the lungs where gas exchange occurs.

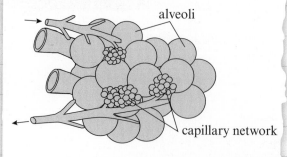

alveoli

capillary network

Describe how the structures shown are adapted to this function. **(4 marks)**

The cell surface membrane

The cell surface membrane contains the cell, controls movements of substances in and out, maintains the osmotic balance of the internal environment and allows the cell to be recognised by, for example, hormones.

The chemical structure of the cell membrane

- lipids (in the form of phospholipids and cholesterol)
- proteins
- carbohydrate (in the form of glycoproteins and glycolipids).

Properties of cell membranes

- flexible
- fluid
- selectively permeable.

Developing models of the cell membrane

To find out how these components are arranged took many years and awaited the development of techniques such as electron microscopy and the use of radioactive isotopes. The resulting data were used to build models such as the **fluid mosaic model** to provide a scientific explanation of the structure and properties of cell membranes.

The fluid mosaic model

outside cell

hydrophilic head　glycolipids

glycoprotein

hydrophilic tail　protein

inside cell　cholesterol　channel　lipid bilayer

The fluid mosaic model has a fluid lipid bilayer with a mosaic of proteins, glycoproteins and glycolipids floating in it.

Facts and evidence for the model

- Phospholipids are both hydrophilic and hydrophobic so form bilayers in an aqueous environment.
- A monolayer film of phospholipids is twice as large as the cell surface area.
- Microscope images of cell surfaces show proteins sticking out.
- When lectins, which react with carbohydrates, are added to a membrane they are found only on the outside.
- Some water-soluble substances pass into and out of cells.
- Ionic and polar molecules do not pass easily through membranes, but lipid-soluble substances do.

Worked example

The Davson–Danielli model (below), proposed in 1935, accounts for protein and lipid, but it does not explain many of the known facts about membranes.

| protein layer |
| phospholipid bilayer |
| protein layer |

Compare and contrast the Davson–Danielli model with the fluid mosaic model. **(3 marks)**

Both have a phospholipid bilayer with protein but the fluid mosaic model has proteins within the phospholipid layer while the Davson–Danielli model has protein layer on the outside of the membrane.

The fluid mosaic model has glycolipid, glycoprotein and cholesterol.

With 'compare and contrast.' questions you are looking for the similarities and differences of two (or more) things. The answer must include at least one similarity and one difference.

Now try this

Describe the structure of a cell membrane. **(4 marks)**

Passive transport

Membranes control the movement of materials in and out of cells and organelles. Substances can move across membranes either **passively** (sometimes with help, but energy is never needed) or **actively**, which needs energy. How exactly they move depends on their properties.

Passive movement

There are three kinds of passive movement: diffusion, facilitated diffusion and osmosis.

In all of these, when there are more particles in one area (high concentration) than another (low concentration), more move away from the high concentration area than from the low concentration area. So the **net movement** (the balance of movement) is **away** from the high concentration **to** the low.

Equilibrium

Particles move across cell membranes in **both directions all the time**. If the concentration is the same on both sides of a membrane, particles are still moving across in both directions but the **net** movement is zero.

Diffusion

The concentration of a particle may be higher on side A than side B, but there will only be movement if:

- the membrane is permeable (pores big enough)
- the particle and/or pore is not charged
- the particle is soluble.

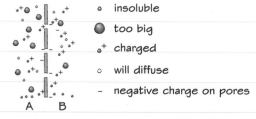

o   insoluble
● too big
o+ charged
o   will diffuse
−   negative charge on pores

A   B

Facilitated diffusion

Small, non-charged molecules, like O_2 and CO_2, can pass through the lipid bilayer. Bigger or polar molecules need a channel in the form of a **channel protein** in order to pass through.

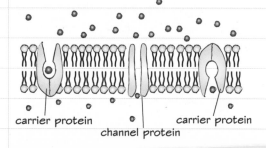

carrier protein   carrier protein
        channel protein

Membrane proteins allowing molecules through a pore (channel protein) or a 'flip-flop' system for specific molecules (carrier protein) causes the protein to change shape, thus carrying the molecule across the membrane.

Osmosis

Osmosis is the diffusion of free water molecules. It involves:

- net movement of water molecules from a solution with low solute concentration to a solution with a higher solute concentration
- movement through a partially permeable membrane (permeable to water but not the solute).

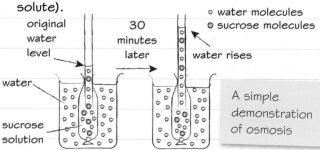

original water level

30 minutes later

o water molecules
o sucrose molecules

water rises

water

sucrose solution

A simple demonstration of osmosis

Now try this

Explain how drugs could be designed to enter a cell rapidly.   **(3 marks)**

Active transport, endocytosis and exocytosis

Living things can move substances in and out of their cells even if the concentration gradient is in the wrong direction (from low to high concentration).

Movement against a concentration gradient

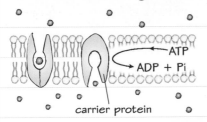

carrier protein

When some channel proteins are given energy (from ATP) they can move molecules **against** a concentration gradient. This is called **active transport**.

Adenosine triphosphate (ATP), the energy currency of the cell

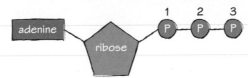

The ATP molecule is made of a ribose, an adenine base and three phosphate groups (P) (see page 103).

When the bond between the third and second phosphate is broken by hydrolysis, energy is released. This can be used in energy-requiring processes taking place within the cell.

Energy is required to add a third phosphate bond to adenosine diphosphate (ADP) to create ATP again.

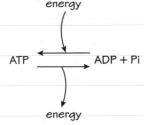

Endocytosis and exocytosis

Vesicles transport large molecules into and out of cells. This is the process of **endocytosis** (movement into the cell) and **exocytosis** (movement out of the cell). Both of these processes are possible because of the fluid nature of the cell membrane.

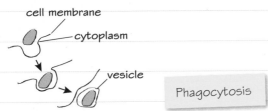

cell membrane
cytoplasm
vesicle

Phagocytosis

Phagocytosis ('cell-eating') is a kind of endocytosis. A large structure, such as a bacterial cell or a molecule, is surrounded by the cell membrane and engulfed in a vesicle, which moves into the cell. If the material being taken in is a liquid it is called 'cell-drinking' (pinocytosis).

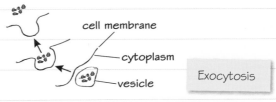

cell membrane
cytoplasm
vesicle

Exocytosis

To remove large particles from inside the cell, the cell performs exocytosis. This is a process in which large molecules in the cell are engulfed in a vesicle. The vesicle then moves to the cell membrane where it fuses with the membrane. The contents that were engulfed are then released out of the cell.

Worked example

Describe **one** similarity and **one** difference between active transport and facilitated diffusion. **(2 marks)**

Both processes can use carrier proteins.

Active transport requires energy whereas facilitated diffusion does not.

Now try this

Explain how insulin, made in the β cells of the pancreas, is moved out of the cell into the pancreatic duct. **(3 marks)**

Practical on membrane structure

 Practical skills Even though an electron microscope is needed to see any detail of a membrane, it is possible to find things about their structure in a simple experiment.

Investigating the effect of temperature on beetroot membrane permeability

The fluid mosaic model of membrane structure suggests that temperature might affect cell membranes. As temperature increases, phospholipids will become more fluid, allowing molecules to leak from the cell. Beetroot cells can be used to test this idea because they contain a pigment called betalain within the vacuole of the cell.

Location of betalain pigment

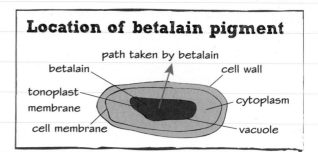

The method

| cut equal sized pieces of tissue | — to control the variable size of sample |

↓

| rinse under running water | — to remove all betalain (pigment) released by cutting |

↓

| place pieces in equal volumes of distilled water:
• use a range of temperatures
• leave for equal time | — temperature is the **independent variable** so must be varied but the volume of the distilled water and the time in it need to be kept constant |

↓

| remove pieces carefully and shake each solution gently | — to disperse any pigment |

↓

| assess amount of pigment lost using a colorimeter to measure the absorbance or transmission value of the solution | — use of a colorimeter ensures the results are quantitative |

↓

| plot values on graph:
• temperature on x-axis
• absorbance or transmission on y-axis | — to allow the relationship between the two variables to be easily seen |

Worked example

Annotate the graph below with an explanation of the shape of the curve.   **(3 marks)**

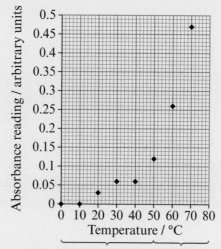

There is little or no pigment leakage between these temperatures because the lipid layer in the membrane is intact.

There is a sudden increase in leakage of pigment out of the beetroot pieces. This is probably due to disruption of the membrane as it becomes more fluid at about 55 °C. Denaturation of membrane protein channels may also play a part.

Maths skills Remember to give an explanation as well as a description of what the graph shows.

Preparing for core practicals

You need to know what you did but also *why* you did it. A flowchart like the one here would be a useful thing to make for all your core practicals.

Now try this

Describe in detail how the dependent variable would be measured in the beetroot experiment.   **(3 marks)**

The structure of DNA and RNA

The instructions in living things are carried by deoxyribonucleic acid (DNA) and ribonucleic acid (RNA).

Structure of mononucleotides

Nucleic acids (DNA and RNA) are **polynucleotides** composed of **mononucleotides**. There are five kinds of mononucleotide, but all have the same basic structure.

phosphate —
pentose sugar —
organic nitrogenc base

The components of a single nucleotide are joined by a **condensation reaction**: the phosphate joins to the sugar at C atom 5 (C5), and the nitrogenous base joins to the sugar at C atom 1 (C1).

Mononucleotides are joined in polynucleotides by **phosphodiester bonds** that are also formed by a condensation reaction.

	DNA	RNA
Pentose sugar	deoxyribose	ribose
Base	adenine (A)	adenine (A)
purines	guanine (G)	guanine (G)
	cytosine (C)	cytosine (C)
pyrimidines	thymine (T)	uracil (U)

Phosphodiester bonds

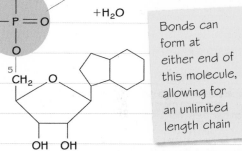

Nucleotides join together when a phosphate group is attached to the C3 of one pentose and the C5 of another. This forms a phosphodiester bond.

phosphodiester bond

$+H_2O$

Bonds can form at either end of this molecule, allowing for an unlimited length chain

Base pairing

DNA contains two separate polynucleotide strands. The strands are joined together by hydrogen bonds between the base of a nucleotide in one strand and a base of a nucleotide in the other strand. These bases form a **complementary base pair** forming a double helix. In complementary base pairing, adenine always pairs with thymine and cytosine always pairs with guanine.

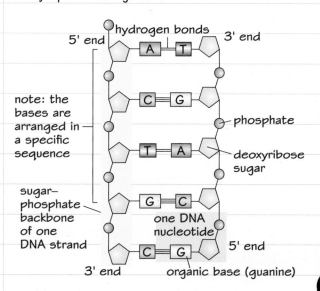

note: the bases are arranged in a specific sequence

5' end   hydrogen bonds   3' end

phosphate

deoxyribose sugar

sugar–phosphate backbone of one DNA strand

one DNA nucleotide

5' end

3' end   organic base (guanine)

In RNA, the base uracil is present instead of thymine, so in RNA adenine pairs with uracil.

The sugar–phosphate backbone

The joining of nucleotides by phosphodiester bonds between the sugar of one nucleotide and the phosphate of another creates a chain of alternating sugar and phosphate groups. This chain is the **sugar–phosphate backbone**. This is the structural framework of polynucleotides such as DNA and RNA.

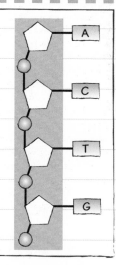

Worked example

Describe two structural features of DNA which give it the ability to replicate.   **(3 marks)**

The molecule is a double strand. The two strands are held together by H— bonds between adenine and thymine or guanine and cytosine. These bonds are easily broken.

Now try this

A section of a DNA molecule has 350 bases. 32% of these are cytosine. Determine how many of each of the four bases are present in this section.   **(3 marks)**

23

Protein synthesis – transcription

A gene is a sequence of bases on a DNA molecule that codes for a sequence of amino acids in a polypeptide chain.

Making a protein

The DNA sequence is copied on to messenger RNA, mRNA, in the nucleus (**transcription**).

The mRNA moves into the cytoplasm and attaches to a ribosome. Ribosomes facilitate the joining together of amino acids in the correct sequence in the **translation** stage of protein synthesis. This stage requires transfer (t)RNA.

Messenger RNA (mRNA)

- formed in nucleus from DNA antisense (template) strand
- single stranded
- not usually folded
- carries codons
- attaches to tRNA.

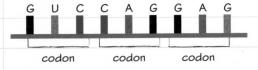

G U C C A G G A G

codon   codon   codon

Be careful! Do not mix up the terms **transcription** and **translation**.

A useful way to think of translation is to remember that going from English to French, for example, is the *translation* from one language to another.

Similarly, going from a base sequence (in mRNA) to amino acid sequence in a polypeptide you are going from one language to another, so it is *translation*.

Worked example

When a gene that contains 22% adenine is transcribed, the mRNA produced will have:

- ☐ **A**   22% adenine
- ☐ **B**   0% cytosine
- ☑ **C**   0% thymine
- ☐ **D**   28% uracil.   **(1 mark)**

This needs some careful thought. Transcription is the making of mRNA on the DNA template strand. There is no thymine in mRNA, but adenine pairs with thymine, you should then realise that it pairs with uracil instead.

Transcription: making an mRNA copy of the DNA template strand

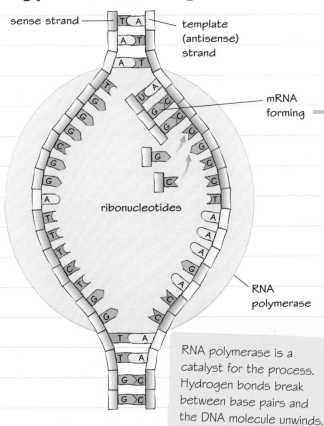

sense strand — template (antisense) strand

mRNA forming

ribonucleotides

RNA polymerase

RNA polymerase is a catalyst for the process. Hydrogen bonds break between base pairs and the DNA molecule unwinds.

Transfer RNA (tRNA)

- single stranded
- folded into specific pattern
- carries anticodons complementary to the codons on mRNA and hence DNA sense strand
- attaches to mRNA and amino acids.

amino acid attachment site

hydrogen bonds

anticodon

tRNA, individual nucleotides not shown

Now try this

Write out the mRNA transcript of the DNA antisense sequence TACTTCCTTTGACGA.

(2 marks)

Translation and the genetic code

Messenger RNA carries part of the genetic code from DNA into the cytoplasm. It is then translated into polypeptides, which happens on ribosomes via transfer RNA.

Four steps to creating a protein

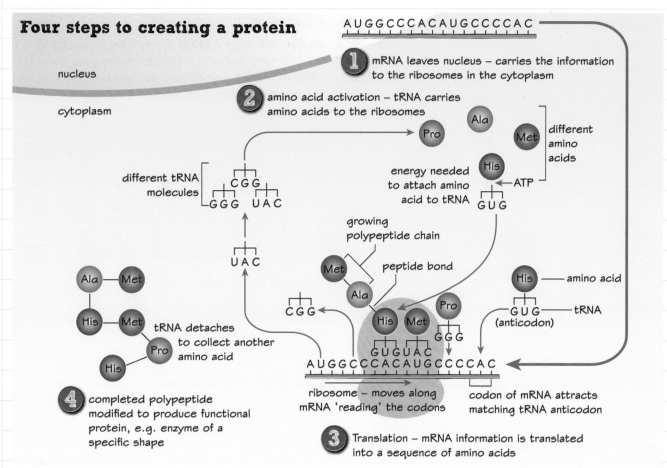

AUGGCCCACAUGCCCAC

1 mRNA leaves nucleus – carries the information to the ribosomes in the cytoplasm

2 amino acid activation – tRNA carries amino acids to the ribosomes

nucleus

cytoplasm

different tRNA molecules

C G G
G G G U A C

Ala

Pro

Met   different amino acids

His

energy needed to attach amino acid to tRNA

ATP

G U G

U A C

Ala — Met

His — Met

His   Pro

His

growing polypeptide chain

Met

Ala

peptide bond

C G G

tRNA detaches to collect another amino acid

His   Met

G U G U A C

His

amino acid

G U G   tRNA
(anticodon)

Pro

G G G

AUGGCCCACAUGCCCAC

ribosome – moves along mRNA 'reading' the codons

codon of mRNA attracts matching tRNA anticodon

4 completed polypeptide modified to produce functional protein, e.g. enzyme of a specific shape

3 Translation – mRNA information is translated into a sequence of amino acids

Genetic code

- The genetic code carried by DNA is a **codon** (three-base code).
- Each codon codes for one amino acid.
- There is one triplet **start codon**.
- There are three **stop codons** (chain terminators).
- The genetic code is non-overlapping; for example, the six bases UCUUUG consists of only two triplet codons, UCU and UUG.
- Transcription of the DNA strand (antisense strand) creates mRNA where the triplet codes are written as RNA bases (A, G, C and U).

Degenerate code

There are 64 possible three-letter combinations if four DNA bases are used as triplets. As there are only 20 amino acids in nature, this means that most amino acids are coded for by more than one triplet and the genetic code is said to be **degenerate**. This means that mutations that alter one of the bases in a triplet code would have no effect on the amino acid incorporated **if** the altered codon codes for the same amino acid.

section of DNA | A | A | T | A | A | C | C | A | G | T | T | T |

amino acids   leucine   leucine   valine   lysine

Worked example

The diagram opposite shows the base sequence of a section of DNA. Using this diagram explain the meaning of the statement 'the genetic code is both non-overlapping and degenerate'. **(4 marks)**

Non-overlapping: each triplet is discrete and adjacent so in this case AAT, AAC, CAG and TTT give four distinct codes.

Degenerate: more than one code can be used for a particular amino acid or stop code.

For example, in the sequence opposite, AAT and AAC both code for leucine.

Now try this

Describe the process of translating mRNA into a protein X.   **(4 marks)**

Amino acids and polypeptides

Proteins have a huge range of structures and functions in the living world, but they are all made of chains of amino acids.

The structure of amino acids

All amino acids share a common structure:

- a central carbon atom joined to a hydrogen
- an amino group (NH_2)
- a carboxyl group (COOH)
- the fourth bond is to one of 20 different (R) groups.

The general structure of an amino acid

Peptide bonds

Amino acids can join together via a **peptide bond** to form a dipeptide. The carboxyl group of one amino acid joins to the amino group of another to form the peptide bond. This happens in a **condensation reaction** in which a molecule of water is lost.

The formation of a peptide bond between two amino acids to form a dipeptide

condensation reaction

H_2O

peptide bond

The formation of polypeptides and proteins

Dipeptides have a carboxyl group at one end and an amino group at the other. This means they can join to more amino acids at either end to form a single chain, which is a **polypeptide**.

With 20 different amino acids and chains hundreds of amino acids long, the variety of possible protein molecules is vast.

Primary structure

The order of amino acids in a protein is called the **primary structure**.

cysteine	alanine	lysine	glycine	leucine

Part of the primary structure of a protein showing five amino acids

Worked example

The diagram shows the R groups of the amino acids tryptophan and phenylalanine.

R group of phenylalanine

$H-C-H$

R group of tryptophan

CH_2-

Draw a molecule of the amino acid phenylalanine. **(3 marks)**

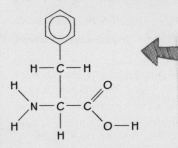

All amino acids have the same structure on three of the four carbon bonds, so all you need to do is remember that and then add the R group in the correct place.

Now try this

Proteins are polymers of amino acids joined by peptide bonds formed between the:

A R groups

B R group and the amino group

C R group and the carboxyl group

D carboxyl group and the amino group. **(1 mark)**

Folding of proteins

When a polypeptide folds, a fully functional protein is produced. The pattern of this folding depends on the sequence of amino acids or primary structure (you learned about primary structure of a protein on page 26). The first level of folding is the secondary structure and then further folding is the tertiary structure.

Secondary structure – four major facts

1 Chains of polypeptides fold up to form proteins.

2 Amino and carboxylic acid groups in the amino acid chain carry small amounts of charge.

3 The charge is negative (δ^-) on the CO of the carboxyl group and positive (δ^+) on the NH of the amino group.

4 These charges result in **hydrogen bonds** forming between parts of the chain, which stabilise the structure.

Hydrogen bonds

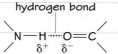
hydrogen bond
N—H⋯⋯O=C
δ^+ δ^-

A hydrogen bond between two amino acids in a polypeptide chain. δ (delta) means small charge, + or −.

Secondary structure – two types

The secondary structure of a protein is held together by many hydrogen bonds, to form either an α-helix or a β-pleated sheet:

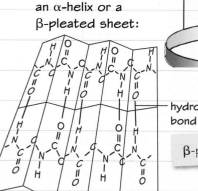

α-helix

hydrogen bond

hydrogen bond

β-pleated sheet

Tertiary structure

The α-helix or β-pleated sheet structures fold further to give the unique **tertiary structure** of a protein. This structure is held together by hydrogen bonds, ionic bonds and disulfide bonds, all between R groups.

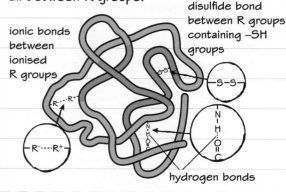

ionic bonds between ionised R groups

disulfide bond between R groups containing –SH groups

hydrogen bonds

Quaternary structure

In some proteins, a number of folded chains join together to form the **quaternary structure**. Each folded chain is joined to the others by the same bonds as in the tertiary structure.

polypeptide A

polypeptide B

Worked example

The diagram shows an insulin molecule.

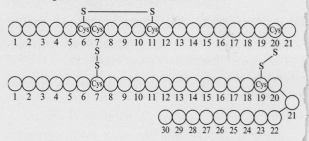

Explain how the sequence of amino acids determines the shape of a protein with reference to insulin.

(3 marks)

Amino acids have different R groups. This determines which bonds can form between them in the chain. For example, if there are cysteines then S–S bonds can form between them, such as between insulin chain A amino acid 20 and chain B amino acid 19.

If you are asked to make reference to a particular example then make sure you do! The example might be expected from your knowledge, or it might (as here) be given in the stem.

Now try this

Which of the listed bonds could hold together the tertiary structure of a protein?
Peptide, hydrogen, ionic, ester, disulfide, glycosidic. **(3 marks)**

Haemoglobin and collagen

The structure of a protein determines its properties and therefore its function. This applies to: enzymes, antibodies, muscle proteins, hormones, structural proteins, transport proteins, and many others.

Two examples of proteins with quaternary structure

There are two major categories of protein – fibrous (e.g. collagen) and globular (e.g. haemoglobin).

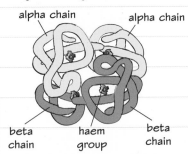

alpha chain   alpha chain

beta chain   haem group   beta chain

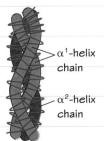

α^1-helix chain

α^2-helix chain

Haemoglobin (Hb) is a globular protein, with four polypeptides (two α and two β) and four haem groups. The haem groups are **prosthetic groups**.

Collagen is a fibrous protein with three polypeptide chains (two α_1 and one α_2, where α_1 and α_2 refer to the two helices).

Haemoglobin (Hb)

Structure	Function
globular	can be soluble
polypeptide joined to a haem group which contains iron	haem groups bind to oxygen
four polypeptide & haem groups	binds to four times more oxygen than one haem group would

Cooperative binding

When an oxygen binds to one haem group this makes it easier for the next one to bind due to the changes it causes in the shape (**tertiary structure**). This is **cooperative binding**.

 hard → easier →   easier still →

The structure and function of collagen

Collagen is a strong fibrous protein, which means it is insoluble. The three polypeptide chains are α-helices, and wind around each other to form a rope-like strand held together by hydrogen bonds.

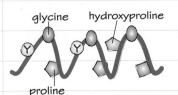

glycine   hydroxyproline

proline

Part of a collagen helix, X and Y are other amino acids

In the three helices, every third amino acid is glycine whose R side chain is very small; it is just −H. This means that it can fit into the very small space on the inside of the triple helix. On either side of this are proline and hydroxyproline, with big R chains. These chains keep out of each other's way to maintain the strong, fibrous structure.

The 3 helices are cross-linked; the strands are staggered, so there are no weak points.

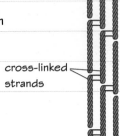

cross-linked strands

Describe structural and functional differences between fibrous proteins and globular proteins.   **(4 marks)**

Fibrous proteins have straight chains whereas globular proteins are compact and often spherical.

Globular proteins are soluble but fibrous proteins are not.

Fibrous proteins are involved in structural functions and globular are not.

Globular are involved in catalysis as enzymes, fibrous are not.

The question asks for both structural and functional differences so you must include both.

With reference to haemoglobin, explain the meanings of the terms quaternary structure and prosthetic group.   **(5 marks)**

Enzymes

Living organisms are chemical processing factories and all the reactions are speeded up (catalysed) by enzymes, which are proteins.

Enzyme structure

Enzymes are globular proteins and all work by binding to their individual **substrate(s)** (the substance(s) on which they act) via an **active site**. This means they must have a very precise 3D shape to fit their particular substrate. This comes from their folding pattern as a globular protein (see page 28).

Inside and outside cells

Enzymes are made inside cells. Some stay there and catalyse reactions within the cell; these are **intracellular enzymes**. DNA polymerase and DNA ligase (page 34) are examples. Some are secreted and work outside the cell; these are **extracellular enzymes**. All of the digestive enzymes are examples of extracellular enzymes.

Lock-and-key hypothesis

Enzymes are highly specific as to what substrate they will work on because of their precise 3D structure. This is known as the **lock-and-key hypothesis**.

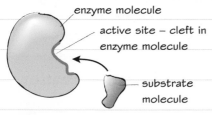
enzyme molecule
active site – cleft in enzyme molecule
substrate molecule

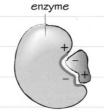
enzyme

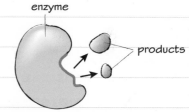
enzyme
products

1 Random movement causes the enzyme and substrate to collide, and the substrate enters the active site.

2 Enzyme–substrate complex forms; charged groups attract, distorting the substrate and aiding bond breakage or formation.

3 Products are released from the active site leaving the enzyme unchanged and ready to accept another substrate molecule.

Induced fit hypothesis

In this refinement of the lock-and-key hypothesis, the enzyme changes shape to fit the substrate when the substrate is near.

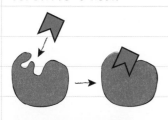

Worked example

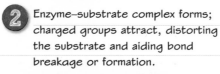

Explain why an enzyme that breaks condensation bonds in starch will not break condensation bonds in a protein. **(4 marks)**

The bonds in starch are glycosidic, whilst those in a protein are peptide.

The enzyme molecule has a specific shape, creating an active site.

The substrate (starch) binds to this site.

This is because, as it nears it, the enzyme molecule changes shape.

The substrate is said to have induced the enzyme to fit it.

A potential substrate with the wrong features, such as a protein, would not cause this induced fit.

Now try this

The diagram shows the breakdown of sucrose by the enzyme sucrase.

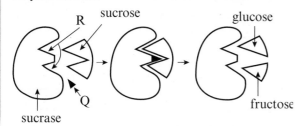
R
sucrose
glucose
sucrase
Q
fructose

(a) Name Q and R.
(b) What class of substance is sucrose in this reaction? **(1 mark)**
(c) What kind of bond is split? **(1 mark)**
(d) What kind of reaction is this? **(1 mark)**
(You may need to refer to page 10) **(1 mark)**

Activation energy and catalysts

Enzymes are **biological catalysts**. They cannot cause a reaction to happen, they simply catalyse reactions that are already happening. They do this by reducing the activation energy .

Breaking bonds

To start a reaction, bonds need to be broken or made, but only a certain number of molecules in a mixture have enough energy to do this.

Reactions can be accelerated by either adding energy or lowering the amount of energy needed.

Heating is a way of adding energy, so heating a reaction mixture can speed up a reaction. However, this cannot be done in living things since the heat would damage their cells, so the amount of energy needed is lowered. Enzymes work by lowering this energy requirement.

Activation energy

Enzymes lower the **activation energy** needed to make the reaction happen, so that it occurs more easily.

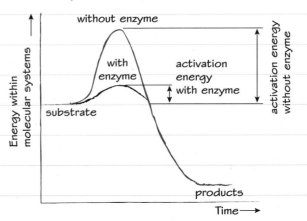

More molecules can react when the enzyme is present

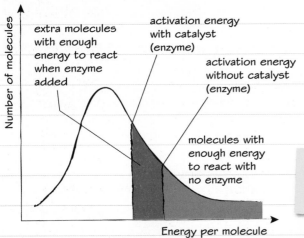

How do enzymes lower activation energy?

- They provide an alternative, lower energy, pathway.
- They provide a more favourable pH in the **active site.**
- They put strain on bonds in the substrate to break them.
- They bring reactants close together so bonds are easier to form.

The enzyme provides an active site where reactions occur more easily, lowering the activation energy required, which means that more molecules have enough energy to react.

Worked example

Explain why living organisms use enzymes.   **(3 marks)**

Enzymes provide a lower energy pathway to convert substrates into products.

They do this by binding with the substrate, and providing an environment that makes the reaction more likely to occur.

This allows reactions to occur without the need to raise the temperature of the cell.

The answers to explain questions must contain some element of reasoning or justification. Good words to use are **because** and **therefore**, amongst others.

Now try this

Explain the meaning of the terms biological catalyst, activation energy and active site.   **(5 marks)**

Reaction rates

 Practical skills The rate of an enzyme catalysed reaction slows down as substrate is used up, so to get a true measure of reaction rate, the **initial rate** should be measured. The initial rate is the rate before the concentration of substrate starts to become limiting. Only if this is measured, can rates be compared.

▦ Maths skills   Calculating the initial rate of a reaction

| protease is used to break down a cloudy protein solution |
↓
| decrease in cloudiness is measured with a colorimeter, taking an absorbance reading every few seconds |
↓
| results can be plotted as absorbance (y) against time (x) |
↓
| the gradient of the straight line graph is calculated |
↓
| from this, the initial rate is calculated |

Protease experiment at enzyme concentration of 1%

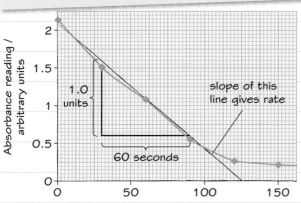

- For this reaction, the rate is the change in absorbance over time.
- You can calculate the rate using the graph by finding the gradient, $\frac{\text{change in y-values}}{\text{change in x-values}}$
- Draw a **straight line of best fit** through the initial part of the graph (before the curve changes due to limited substrate).
- The change is 1.0 absorbance units in 60 seconds 1 unit min^{-1} at 1% enzyme concentration.

The effect of enzyme concentration

- Repeating the experiment with other enzyme concentrations gives initial rates for each concentration.
- Plotting these results gives a graph showing the effect of enzyme concentration on initial reaction rate.

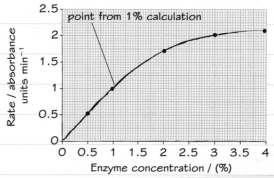

Effect of enzyme concentration on initial reaction rate. In this case, substrate concentration becomes limiting. If substrate were in excess the graph would be a straight line.

True rate

- The experiment on this page does not actually give a true rate as the units are arbitrary (AU).
- To convert to a true rate you would need to convert absorbance to mg of protein. You would do this by calibrating the colorimeter with different concentrations of protein solution.

Be careful! A time taken is **not** a rate! $\frac{1}{\text{time}}$ gives an indication of rate, but it is still not a rate without knowing how much substrate is used or how much product is produced.

Worked example

Explain the effect of enzyme concentration on initial reaction rate when substrate is in **excess**. **(4 marks)**

As concentration of enzyme increases, so does initial rate of the reaction.

Because the substrate is in excess, it is not limiting so rate of reaction would continue to increase and the graph would be a straight line.

Very often when you are asked to explain something there will need to be some description as well.

Now try this

Explain why it is necessary to measure the initial rate of reaction when investigating the effect of enzyme concentration on the rate of reaction.

(3 marks)

Initial rates of reaction

🧪 **Practical skills** Investigate the effect of enzyme and substrate concentrations on the initial rates of reactions. The **dependent variable** (DV) is the variable to be measured or observed and is the initial rate of reaction. The **independent variable** (IV) and is the variable that will be changed in a controlled way, and this will be either enzyme concentration or substrate concentration.

Factors that affect enzymes and how to control them

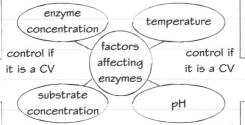

enzyme concentration; should be in excess if substrate concentration is the IV	use a water bath; the value should be in the correct range for the enzyme used, as near to the optimum as possible

control if it is a CV

control if it is a CV

substrate concentration; should be in excess if substrate concentration is the IV	pH; use buffer solutions, again in the correct range for the enzyme used, as near to the optimum as possible

control if it is a CV

control if it is a CV

Skills covered in Core practical 4

- Using appropriate apparatus, record the **control variables** (CVs, the variables that you need to keep the same throughout the experiment), e.g. mass, time, volume, temperature and pH.

- Using a colorimeter if enzyme reaction affects absorbance; when peroxidase catalyses the breakdown of hydrogen peroxide to oxygen and water, the volume of oxygen produced is measured.

- Using laboratory glassware.

- Using data logger to collect data and using software to process data.

- Considering health and safety issues.

🖩 **Maths skills** Calculating gradients from lines is a Maths Skill you need. You should draw a line of best fit through the straight line part of the graph to calculate the average gradient rather than the gradient at one particular point.

Worked example

Lipase can be used in the manufacture of biodiesel. A series of experiments have been carried out to determine the effect of lipase concentration on the rate of the reaction that creates biodiesel. The graph shows the course of the reaction at a given enzyme concentration. The table shows initial rates for all the lipase concentrations in the experiment.

For which enzyme concentration has the graph been plotted? Explain your answer. **(4 marks)**

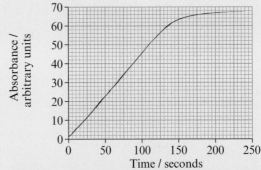

Enzyme concentration (%)	Initial rate of reaction / arbitrary units s⁻¹
0.0	0.00
0.5	0.20
1.5	0.44
3.0	0.75
4.5	0.76
5.5	0.79

The gradient of the graph gives rate of reaction.

Gradient is $\frac{40}{90}$ = 0.44 AU sv⁻¹.

Using the table, the enzyme concentration this graph represents is 1.5%.

Now try this

In the experiment in the Worked Example, state **two** control variables and explain how they could be controlled. **(4 marks)**

Exam skills

This exam-style question uses knowledge and skills you have already revised. Have a look at pages 10 and 29 for reminders about carbohydrate chemistry and enzymes.

Worked example

(a) A student carried out an experiment on substrates for respiration in yeast. A healthy yeast culture was incubated with different substrates and the time taken for the yeast to produce $8\,mm^3$ of CO_2 was recorded.

Sugar	Time in seconds to produce $8\,mm^3$ CO_2
glucose	85
lactose	no CO_2 produced
sucrose	180
ribose	no CO_2 produced

Calculate the percentage increase in the rate of production of CO_2 when glucose is used compared with sucrose. **(3 marks)**

Rate of reaction with glucose is $\dfrac{8}{85}$ = 0.09

0.09 mm³ per second

Rate with sucrose is $\dfrac{8}{180}$ = 0.04

0.04 mm³ per second

Difference in rates = 0.09 – 0.04 = 0.05 mm³ per second

Percentage increase in rate with glucose compared with

sucrose = $\dfrac{0.05}{0.04}$ × 100

= 125%

(b) Explain the differences found between the carbon dioxide production rates of the four sugars investigated. **(5 marks)**

Glucose has the fastest rate because it is a monosaccharide and therefore does not need to be broken down before it can be respired.

Sucrose is a disaccharide and has to be digested into its two monosaccharide components (a molecule of glucose and one of fructose) before it can be respired and therefore less CO_2 is made per second. Lactose is also a disaccharide but the yeast does not have the enzyme necessary to break it down so cannot use it in respiration. This is because an enzyme which can digest one disaccharide will not necessarily be able to break the bond in a different disaccharide due to its specificity.

Ribose is a 5C sugar and it is likely that the yeast does not have the necessary enzymes to respire it.

Whenever you are told about a study you will be given details of what was done. You must read these **carefully** and assume they are all important to your understanding and for the questions to follow. Keep referring back to this information as you work through questions. If data is given, as here, you will need to use it in your answer.

 Maths skills It is important to set out problems that involve mathematical steps logically, as shown. This helps you and the examiner to see clearly what you have done. Always do a sensible estimate of the answer. A common mistake with percentage calculations is to get the calculation the wrong way round. In this case that would be 0.04 ÷ 0.05, giving 80%. If you did a sensible estimate you would know this cannot be correct as the rate when using glucose (0.09) is more than double the rate of using sucrose (0.04).

Often, as here, **explain** is asking you to come up with some suggestions about what is going on. You are not expected to know the answer but to work it out from the information given and your own knowledge. Here the knowledge you need is about enzyme specificity and carbohydrate structure. You need to link observations of the data to your explanation, with words such as **because** and **therefore**.

DNA replication

A key feature of the genetic material is that it can make identical copies of itself, generation after generation. The genes it carries will then be passed on from parents to offspring.

The mechanism of DNA replication

1 The two strands of DNA unwind and split apart.

2 Free nucleotides line up along each strand, according to complementary base pairing rules.

3 DNA polymerase joins the nucleotides together as a phosphodiester bonds form between each deoxyribose and adjacent phosphate group; DNA ligase joins partly formed strands together.

4 Hydrogen bonding links the two strands together.

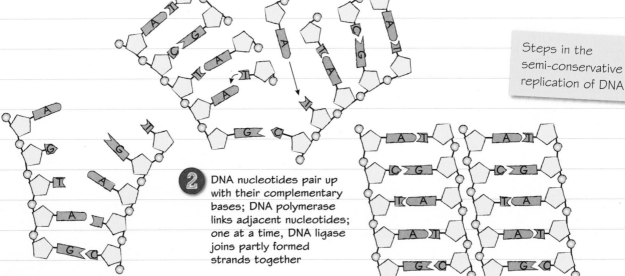

Steps in the semi-conservative replication of DNA

2 DNA nucleotides pair up with their complementary bases; DNA polymerase links adjacent nucleotides; one at a time, DNA ligase joins partly formed strands together

1 hydrogen bonds between the bases break, allowing the DNA to 'unzip'; this is catalysed by DNA helicase

3 two identical daughter strands are created

Worked example

Describe **two** differences between the processes of DNA replication and transcription.　　**(2 marks)**
(See page 24 for a reminder of transcription.)

In DNA replication both strands are copied but in transcription only one strand is.

In replication DNA polymerase is used, in transcription the enzyme involved is RNA polymerase.

If this were a compare and contrast question (which it could have been) you would need to give at least one difference and one similarity. In this one you just need to come up with the differences.

This would get the 2 marks, but there are other differences too. Can you think of them?

Now try this

Describe the process of DNA replication using the strand shown.　　**(5 marks)**

A	T
C	G
T	A
A	T
G	C

Evidence for DNA replication

Watson and Crick suggested in 1953 how DNA might replicate but it was not until the experiments by Meselson and Stahl (below) that their suggestion was shown to be correct.

The evidence for DNA replication

There are three ways DNA could replicate:

1 **conservative** – when one new double helix is made and the old one remains intact

2 **dispersive** – when the new molecules are a mix of new and old parts

3 **semi-conservative** – when each new molecule has an old strand and a new one; this was suggested by Watson and Crick.

Meselson and Stahl showed that the actual method was semi-conservative in experiments in the late 1950s.

Meselson and Stahl's classic experiment, which demonstrated semi-conservative DNA replication

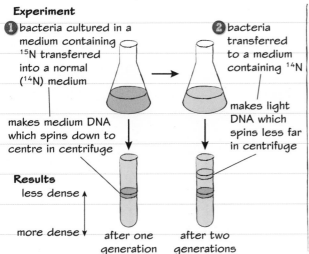

Experiment

1 bacteria cultured in a medium containing ^{15}N transferred into a normal (^{14}N) medium

makes medium DNA which spins down to centre in centrifuge

2 bacteria transferred to a medium containing ^{14}N

makes light DNA which spins less far in centrifuge

Results

less dense ↑

more dense ↓

after one generation   after two generations

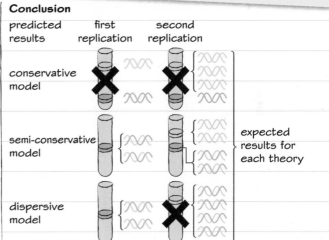

Conclusion

predicted results | first replication | second replication

conservative model

semi-conservative model   expected results for each theory

dispersive model

The graph below shows the amount of DNA present in a sample after one DNA replication.
Complete the graph.   **(2 marks)**

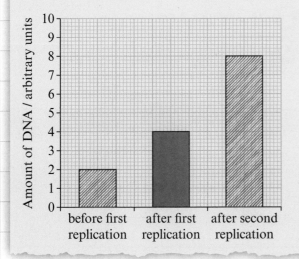

The simple fact you need here is that the amount of DNA doubles each time it replicates. So, if there are 4 AU after one replication, there must have been 2 before it. By the same reasoning, there will be 8 AU after a second replication.

It might seem an obvious point but you must draw bar charts here and not plot points.

Draw a centrifuge tube as it would appear after a third generation of the Meselson and Stahl experiment above.   **(2 marks)**

Mutation

During any stage when the genetic code is copied, mistakes can be made in the new base sequence formed. These mistakes are called **mutations**. Some mutations cause inherited conditions such as **cystic fibrosis**.

Mutations

1. DNA replication creates new cells. In gametes, mutations can be passed to offspring and lead to genetic disorders, such as cystic fibrosis (CF).

2. Transcription creates new mRNA. In body cells, mutations can lead to cancer.

3. Translation creates proteins. Not all mutations of proteins are harmful; many are neutral and do not affect the function of the protein and some can even confer advantages on an organism, leading to evolution.

Mutations that occur during DNA replication can have the greatest effect because they are passed to new cells.

How does the DNA mutation affect a person with CF?

A sticky mucus is produced which affects gas exchange, reproduction and digestion (page 38 tells you more about symptoms).

1. Cl^- is pumped into the cell across the basal membrane.

2. Cl^- diffuses through the open CFTR channels.

3. Na^+ diffuses down the electrical gradient into the mucus.

4. Elevated salt concentration in the mucus draws water out of the cell by osmosis.

5. Water is drawn into the cell by osmosis.

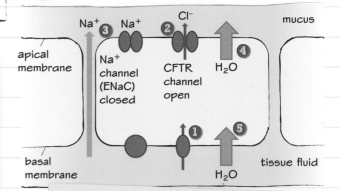

A healthy person's mucus-secretions prevent mucus becoming too viscous when there is too little water in the mucus

1. CFTR channel is absent or not functional.

2. Na^+ channel is permanently open.

3. Water is continually removed from mucus by osmosis.

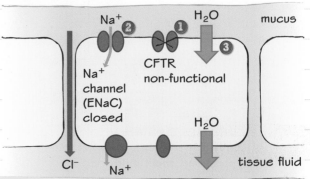

The mucus-secreting cells of a person with cystic fibrosis do not prevent the mucus becoming too viscous when there is too little water in the mucus; the mucus becomes even more sticky and viscous

Worked example

The table shows a number of ways in which the CFTR protein might be affected by a mutation.

For class I cystic fibrosis, explain how a mutation in the CFTR gene could result in no CFTR protein being synthesised. **(3 marks)**

Class	Effect on the CFTR protein
I	CFTR protein is not synthesised.
II	CFTR protein is mis-folded and is not found in the correct location.
III	CFTR protein is mis-folded and is found in the correct location, but does not function properly.
IV	CFTR protein has a faulty opening.
V	CFTR protein is synthesised in smaller quantities than normal.
VI	CFTR protein breaks down quickly after it is synthesised.

The mutation changes the sequence of bases.

This might lead to a stop codon being found before the whole protein has been made.

Therefore the protein is too short and serves no function.

Now try this

Explain the problems that will be faced by a person with any of class I to VI effects. **(4 marks)**

Classical genetics

Some simple laws explain inheritance in all living things. They are called **Mendel's Laws** after Gregor Mendel, who carried out many breeding experiments from which he derived his laws. Capital letters are used for dominant alleles; lower case letters for recessive alleles.

Monohybrid inheritance

Monohybrid inheritance is the inheritance of one characteristic. One gene controls height in pea plants, and it has two forms or **alleles**, one for tallness and one for dwarfness. The diagram shows a cross between a tall pea plant and a dwarf pea plant.

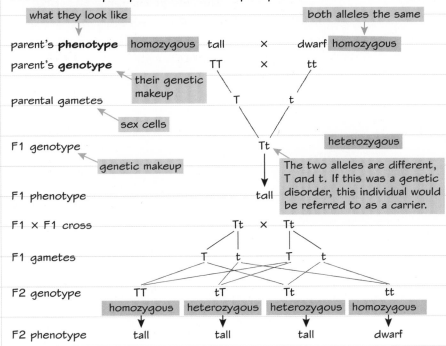

what they look like

both alleles the same

parent's **phenotype** — homozygous   tall   ×   dwarf   homozygous

parent's **genotype** — TT   ×   tt

their genetic makeup

parental gametes — T   t

sex cells

F1 genotype — Tt   heterozygous

genetic makeup

The two alleles are different, T and t. If this was a genetic disorder, this individual would be referred to as a carrier.

F1 phenotype — tall

F1 × F1 cross — Tt   ×   Tt

F1 gametes — T   t   T   t

F2 genotype — TT   tT   Tt   tt
homozygous   heterozygous   heterozygous   homozygous

F2 phenotype — tall   tall   tall   dwarf

Definitions

Gene: a sequence of bases on a DNA molecule that codes for a sequence of amino acids in a polypeptide chain; it is the molecular unit of heredity.

Allele: a form of a gene, e.g. 'tall' and 'dwarf' are alleles of the height gene.

Recessive allele: an allele which is not expressed when a dominant allele is present; in peas, 'dwarf' is recessive.

Dominant allele: an allele which is always expressed; in peas, 'tall' is dominant.

Incomplete dominance: neither allele is dominant and the resultant phenotype is a mix; e.g. crossing a white snapdragon with a red one, will give pink.

Heterozygous: one of each allele.

Pedigree diagrams

In human genetics, crosses cannot be set up as an experiment so patterns of inheritance are studied by looking at natural crosses (the production of offspring) using genetic **pedigree diagrams**.

Deduce with reasons the genotypes of the offspring A, B, C, D, E and F in the pedigree diagram of a human family.   **(5 marks)**

F: Albinism is recessive so must be aa.

B and C: Must be Aa as they have produced offspring F but they are both normal (they are carriers of albinism).

D and E: Cannot be deduced beyond Aa or AA, as the diagram does not have enough information about them.

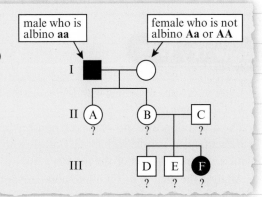

male who is albino **aa**

female who is not albino **Aa** or **AA**

Tay-Sachs disease is a genetic disorder that is inherited recessively. Jen and Adrian are unaffected. Sara, one of their daughters, is also unaffected, but Lucy, another daughter, has Tay-Sachs. Sara has children with Pete, who is unaffected. They have children Dan (Tay-Sachs), Max (unaffected) and Nancy (unaffected). Draw a pedigree for this family.   **(4 marks)**

Cystic fibrosis symptoms

In cystic fibrosis sufferers, a point mutation leads to a missing or faulty protein, which in turn leads to an inability to make runny mucus. This affects the gaseous exchange, digestive and reproductive systems.

The symptoms of CF

Gas exchange

- mucus accumulates in the lungs
- bacteria trapped in mucus increase the possibility of infection
- mucus can block bronchioles
- reduces the number of alveoli in contact with fresh air
- reduces the surface area for gas exchange.

Reproduction

- in men, the vas deferens (sperm duct) is either missing or blocked with mucus, so sperm cannot leave the testes. In women a mucus plug often plugs the cervix so sperm cannot reach the egg.

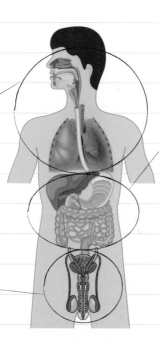

Digestion

- mucus blocks the pancreatic duct
- digestive enzymes cannot reach the duodenum (small intestine)
- food is not properly digested
- leads to tiredness and difficulty in gaining weight
- enzymes trapped within the pancreas cause fibrosed cysts
- damage to insulin producing cells, leading to diabetes.

Worked example

Explain why people with cystic fibrosis are more likely to suffer from lung infections than people without cystic fibrosis. **(4 marks)**

The mucus traps bacteria and other pathogens

and it cannot be removed by the cilia because the mucus is too sticky.

The mucus provides conditions for bacteria to grow and antibodies are not effective.

In addition, the airways can be damaged by trauma caused by coughing.

Now try this

Lung infections in people with CF are caused by bacteria such as *P. aeruginosa* and *S. aureus*.

The graph opposite shows the relationship between the percentage of people with cystic fibrosis who have a lung infection and the age of the person.

Analyse the data in the graph to compare and contrast the relationship between the age of a person and the incidence of a lung infection due to *P. aeruginosa* and *S. aureus*. **(5 marks)**

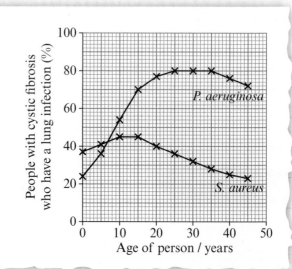

Genetic screening

Prenatal DNA testing of embryonic cells using amniocentesis or chorionic villus sampling (CVS) can be done to determine if the baby has a condition. If the baby has a condition, the parents can decide whether to have an abortion or to continue with the pregnancy and start treatment immediately after the birth.

Amniocentesis

Chorionic villus sampling

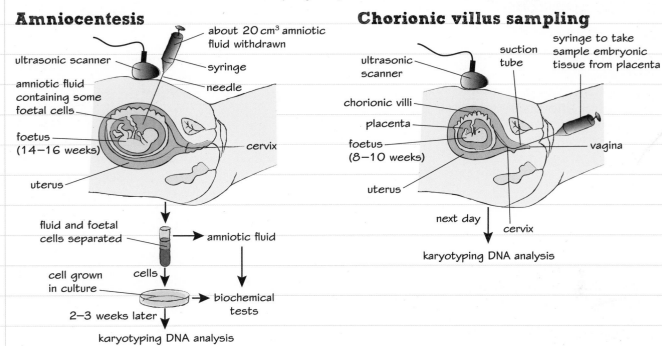

Amniocentesis labels:
- ultrasonic scanner
- about 20 cm³ amniotic fluid withdrawn
- syringe
- needle
- amniotic fluid containing some foetal cells
- foetus (14–16 weeks)
- cervix
- uterus
- fluid and foetal cells separated → amniotic fluid
- cells
- cell grown in culture
- 2–3 weeks later
- biochemical tests
- karyotyping DNA analysis

Chorionic villus sampling labels:
- ultrasonic scanner
- suction tube
- syringe to take sample embryonic tissue from placenta
- chorionic villi
- placenta
- foetus (8–10 weeks)
- vagina
- uterus
- next day
- cervix
- karyotyping DNA analysis

Ethical issues

There is no right decision on what to do about passing on a genetic disorder, but ethical frameworks might help. There are four major frameworks:

- rights and duties
- maximising the good (utilitarianism)
- making decisions for yourself
- leading a virtuous life.

Decisions

Factors to be considered when deciding what is 'best' in relation to prenatal screening are:

- risk of miscarriage or harm to foetus from the tests
- religious beliefs – right to life of the foetus
- potential abortion in the event of a positive diagnosis
- the cost of bringing up a baby that is 'disabled'
- mental and emotional issues surrounding the birth of a 'disabled' baby
- choosing not to become a parent
- dealing with the risk/consequences of false negatives and false positives.

Uses of screening

Identifying carriers

In cases where the condition is recessive (see page 37 for a reminder about recessive alleles) people could be carriers but have no sign of the condition. A sample of blood or cells can be used to detect abnormal alleles in people without the disease who are heterozygous.

Pre-implantation genetic diagnosis (PGD)

Embryos created through *in vitro* fertilisation (IVF) are tested to see if they carry the faulty allele. Only those which have a normal allele are implanted into the woman. This method has the disadvantage that IVF is expensive and quite unreliable. There are also ethical issues surrounding the spare embryos.

Worked example

Discuss one ethical issue relating to the use of prenatal genetic screening. **(2 marks)**

One possible decision after testing is to have an abortion, but all individuals have a right to life and some people think that this includes the foetus, so they believe abortion is always morally wrong.

Now try this

*Evaluate the relative merits of amniocentesis and CVS. **(6 marks)**

Make sure you structure your answer logically, showing how the points you make are related or follow on from each other where appropriate.

Exam skills

This exam-style question uses knowledge and skills you have already revised. Have a look at pages 19 and 22 about membranes.

Worked example

(a) The following shows the structure of a cell membrane.

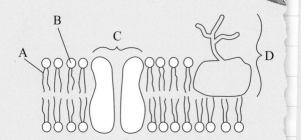

(i) Which letter labels a molecule that is a single polymer?
A, B, C or D

C

(ii) Which letter labels a molecule that is only ever found on the outside of membranes?
A, B, C or D

D

(iii) Which letter labels a hydrophobic fatty acid?
A, B , C or D

A

(b) A student decided to investigate the effect of temperature on the permeability of beetroot cell membranes.
The results are shown in the graph.

Describe how these data could have been obtained. **(5 marks)**

Discs of equal size were cut from beetroots of the same age and variety. The discs were washed to remove any pigment leakage due to cutting. The discs were placed in test tubes containing equal volumes of water. These were placed in water baths at each of the temperatures. They were left for the same amount of time. A colorimeter was used to measure the degree of redness.

(c) Analyse the data from the experiment to explain the effect of temperature on beetroot membrane permeability. **(5 marks)**

The membrane is not permeable to the pigment molecules that cause the red colour, up to 40°C. Above this temperature, the membrane becomes rapidly more permeable.

Above 40°C, the membrane begins to break down due to an increase of kinetic energy of the phospholipids in the membrane. Proteins in the membrane are denatured. Both of these allow diffusion of pigment molecules through the membrane and out of the cells. Above 60°C the concentration of pigment inside the cells is equal to that outside so an equilibrium is reached and the degree of redness remains unchanged at temperatures higher than 60°C.

Synoptic questions: you might at first sight think this going to be a question about membranes. Some of it is, but not all of it. Some parts of the exam can test any material from the units it covers (in this case we are assuming that is topics 1–4). Part (i) is testing that you know that C is a protein (from membrane structure) but also that proteins are made of monomers and are therefore polymers. Part (iii) tests basic lipid biochemistry.

Practical skills

As with all core practicals (this is CP5) you are expected to know how things are done and why they are done in this way. This CP is very good at getting you to think about the control of non-experimental variables.

Assessment Objective 3 (AO3) expects you to 'Analyse and interpret scientific evidence to make judgements and reach conclusions'. This leads to questions like this, which ask you to **analyse** data to **explain** the effect. In this case, work systematically through each part of the graph to explain what you think is happening and how it might be explained.

Exam skills

This exam-style question uses knowledge and skills you have already revised. Have a look at pages 24, 36, and 37.

Worked example

1 The bases found on the anticodons of some of the tRNA molecules coming, in succession, to an mRNA during the synthesis of a protein were analysed. It was found there was 20% adenine, 54% cytosine, 26% guanine, 0% thymine and 36% uracil.

   (a) Deduce the percentage of each base that would be found on the DNA template strand on which the mRNA was made.   **(1 mark)**

   (a) There would be 20% adenine, 54% cytosine, 26% guanine, 36% thymine and 0% uracil.

   (b) Explain your answer.   **(3 marks)**

   (b) The template strand of the DNA carries complementary base pairs to the mRNA made from it, with the exception of thymine which is substituted by uracil in all RNAs. Thus, the DNA carries the same bases as the tRNAs which complement the mRNA, again with the exception of uracil.

2 Achondroplasia is an inherited form of restricted growth in humans. It is caused by a dominant allele. Individuals homozygous for the allele for achondroplasia are rarely born alive.
   Calculate the probability, as a percentage, of a child inheriting achondroplasia if the mother is heterozygous for achondroplasia and the father has normal growth.   **(3 marks)**

A represents the achondroplasia allele and a the normal height allele.

The mother is Aa and the father is aa.

So their gametes would be A and a for her and a for him.

This means the F1 genotypes would be Aa (achondroplastic) and aa (normal growth) in the ratio 1 : 1. This means the probability of a child inheriting achondroplasia is 50%.

3 *Explain how a gene mutation may result in a protein that does not function normally.   **(6 marks)**

A gene mutation is a change in the base sequence in a DNA molecule. This causes a change in the mRNA base sequence, so a different amino acid could be included in the protein. The protein will have a different primary structure from the one it should have. The functionality of the protein depends on its 3D shape, which in turn depends on the R groups on the amino acids being in the correct sequence. If they are not, the bonding in the protein could be different, leading to a change in shape and loss of normal functionality.

It is very important to read the stem of questions carefully. It would be very easy in this case to misread the kind of RNA being discussed, and if read as mRNA (which is more usual for this type of question) you could very easily become confused.

You should always show your working. In this case, the working is unusual. It could be set out as shown with the details listed, or as a genetic cross diagram as below.

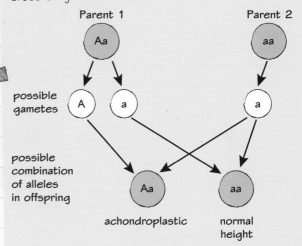

You should always show your working. In this case, the working is unusual. It could be set out as shown with the details listed, or as a genetic cross diagram as below.

Parent 1   Aa   Parent 2   aa

possible gametes   A   a   a

possible combination of alleles in offspring   Aa   aa

achondroplastic   normal height

The * here indicates that marks will be awarded for your ability to structure your answer logically: the points that you make should be related and follow on from each other. Here there is a clear logical sequence from the mutation to the eventual non-functional protein. You could use a flow chart to answer a question like this.

Make sure you also use scientific terminology.

An example would be a disulfide bridge not being formed where it should be.

Prokaryotes

Prokaryotes are very different from **eukaryotes**. They are single-celled and have no membrane-bound organelles.

Prokaryote cell structure

Bacteria and archaea are prokaryotes. This page will focus on bacteria.

ribosomes
The 'workbenches' upon which 70S proteins are made (page 25).

plasmid
Plasmids are double-stranded DNA in a circular structure. They often contain additional genes that aid the bacterium's survival, such as antibiotic resistance or toxin producing genes.

capsule

pili
Thin, protein tubes, which allow bacteria to stick to surfaces. Not found in all bacteria.

cytoplasm

nucleoid
Prokaryotic cells have a single circular length of DNA, carrying all the essential information. The DNA is folded in a region known as the nucleoid.

cell surface membrane

cell wall
In all bacteria, the wall is made of **peptidoglycan**, a polymer of a sugar (-glycan) and some amino acids (peptido-).

flagellum
A whip-like structure which helps in movement and can be a sensory structure. It is not found in all bacteria.

Mesosomes

Mesosomes are infolded in the plasma membrane of the bacterial cell wall. They have been seen in micrographs, but now most scientists think they are **artefacts** of the preparation procedure, so they don't actually exist in living bacteria!

70S ribosomes

In prokaryotes **ribosomes** are size 70S (S is a 'Svedberg', a measure of size by rate of sedimentation). They are made up of a 30S and a 50S subunit.

30S + 50S → 50S / 30S } 70S

The values for the individual subunits don't add up to the value for the whole ribosome, since the rate of sedimentation is related in a complex way to the mass and shape of the molecule.

Worked example

Label the bacteria in the diagram using the descriptions below.

Bacterium P has a single flagellum to enable it to move whilst bacterium Q has several flagella.

Only bacterium R has visible plasmids and bacterium S has an infolding of its cell surface membrane.

Bacterium T has a slime capsule.   **(4 marks)**

 Bacterium ...R....

 Bacterium ...S.....

 Bacterium ...P.........

 Bacterium ...T.....

 Bacterium ...Q.....

Now try this

The diagram shows a mitochondrion, which has much in common with prokaryotes.

Which lettered features are also found in prokaryotes? Explain your answer.   **(3 marks)**

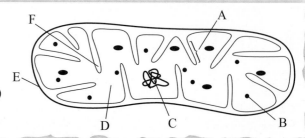

Eukaryotes

All living things are made of cells. All plants, animals and fungi are **eukaryotic** cells.

Cells are tiny!

It was not realised that living things are made from cells until Robert Hooke invented the microscope in 1665 and saw cells for the first time in tiny slivers of cork.

Light microscopes are very limited as to what they can allow us to see inside cells, because the wavelength of light is the limiting factor.

Electron beams have a much shorter wavelength than beams of light and allow much more detail to be seen through an electron microscope. However, electron microscopes can allow only dead material to be examined.

Common features of cells

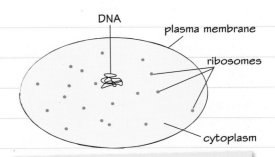

DNA · plasma membrane · ribosomes · cytoplasm

All cells have the four features shown, although they are not always the same.

The ultrastructure of eukaryotic cells

Eukaryotic cells contain **organelles**, which are structures in cells with specialised functions often bound by a membrane.

rER (rough endoplasmic reticulum)
- a series of single, flattened sacs enclosed by a membrane
- ribosomes on the surface proteins made here

nucleolus
- dark staining area within the nuclear envelope
- region of dense DNA and protein
- makes ribosomes

nucleus
- surrounded by a double membrane (envelope)
- pores (holes) in the nuclear envelope

centrioles
- two hollow cylinders
- arranged at right-angles to each other
- makes the spindle in cell division

sER (smooth endoplasmic reticulum)
- a series of single, tubular sacs made of membrane

lipids made here

lysosome
- enclosed by a single membrane
- containing digestive enzymes
- destroys old organelles and pathogens

mitochondrion
- surrounded by a double membrane (envelope)
- inner membrane folded into finger-like projections called cristae (singular, crista)
- central area contains a jelly called the matrix
- containing:
  70S ribosomes
  DNA
- site of respiration

80S ribosomes
site of protein synthesis, contrast with 70S in prokaryotes

Golgi apparatus
- a series of single, curved sacs enclosed by a membrane
- many vesicles cluster around the Golgi apparatus
- modifies proteins and packages them in vesicles for transport

The organelles found in an animal cell, and their functions

Worked example

Which organelle has another organelle inside it? **(1 mark)**

The nucleus

The nucleolus is also an organelle.

Now try this

Make a table listing animal cell organelles with no membrane, those with a single membrane and those with a double. **(3 marks)**

Electron micrographs

We know about the organelles on page 43 because scientists have looked at cells through the electron microscope (EM). It is important to be able to interpret what you see.

Electron micrographs of the key organelles

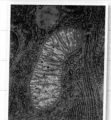

Mitochondrion – double outer membrane and many folded inner ones

Nucleus and nucleolus – nucleus large organelle with double envelope and pores, nucleolus within nucleus and differently stained

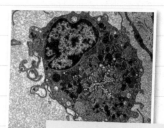

Lysosome – single smooth membrane, dark stained

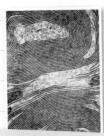

Rough ER – many parallel sacs with ribosomes (smooth ER same with no ribosomes)

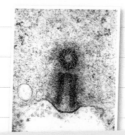

Centrioles – always in a pair with one at right angles to the other

Golgi apparatus – banana shaped sacs parallel to each other

Worked example

The photograph shows an electron micrograph of part of a cell.

(a) Name structure P.   **(1 mark)**

Golgi apparatus

(b) Calculate the width of P at its widest part (between A and B).   **(3 marks)**

$$16\,000 = \frac{16}{\text{width}}$$

Rearrange width $= \dfrac{16}{16\,000}$

$$= 0.001 \text{ mm}$$

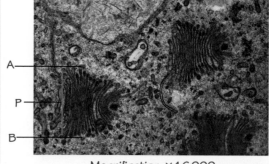

Magnification ×16000

Maths skills   Magnification $= \dfrac{\text{size of image}}{\text{size of real object}}$

As long as you know two of the variables, you can **rearrange the formula** to find the unknown variable.

Now try this

Describe the function of structure P in the photograph in the worked example.   **(2 marks)**

Protein folding, modification and packaging

Once a protein has been made it needs to fold; sometimes other molecules such as carbohydrates are added to it. It is then put in a vesicle if it is going to be secreted (extracellular enzymes for example). All this allows the protein to do its job, for example as an enzyme or structural protein.

Protein trafficking

This is conducted by the rough endoplasmic reticulum (rER) and the Golgi apparatus.

> The pathway of amino acids from incorporation in protein to secretion from the cell. This is called **protein trafficking**. Once made into chains on the ribosomes, the amino acids are folded in the rER and then modified in the Golgi apparatus.

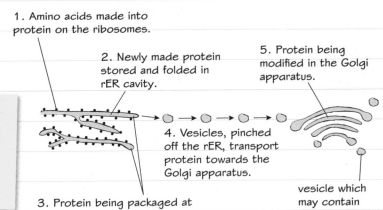

1. Amino acids made into protein on the ribosomes.

2. Newly made protein stored and folded in rER cavity.

5. Protein being modified in the Golgi apparatus.

4. Vesicles, pinched off the rER, transport protein towards the Golgi apparatus.

3. Protein being packaged at the ends of rER. Membrane closes forming a vesicle.

vesicle which may contain extracellular enzymes

Worked example

In a pulse-chase experiment, radioactive amino acids were supplied to pancreatic cells (the pulse). Straight after, a large dose of non-radioactive amino acids were supplied (the chase). The quantity of radioactive amino acid in different parts of the cell was monitored over time. The results are shown below.

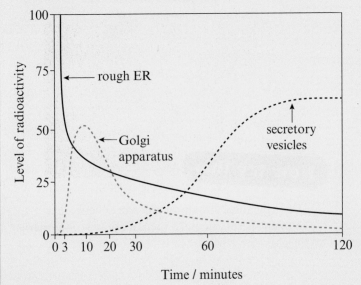

(a) Analyse and interpret the evidence. **(4 marks)**

Radioactivity falls rapidly in the rough endoplasmic reticulum, which shows that amino acids enter and subsequently leave.

 Other data tells us they are in the form of proteins.

As rER radioactivity falls, Golgi apparatus radioactivity increases. This is evidence for movement from rER to Golgi. Finally, as Golgi radioactivity falls, it builds up in secretory vesicles. This shows the modified packaged protein is taken to the cell membrane for secretion.

(b) The quantity of radioactivity in the protein that entered the Golgi apparatus was less than that supplied to the cell. Give **three** reasons for this difference. **(3 marks)**

Some amino acids do not enter the cell and some amino acids are not used in protein synthesis. Some of the protein is elsewhere in the cell.

Now try this

Explain the roles of rough endoplasmic reticulum and the Golgi apparatus in a cell. **(5 marks)**

Sperm and eggs

Mammalian gametes (the **sperm** cell and **egg** cell) are specialised for their functions.

Structure and function in sperm and egg cells

flagellum for movement to swim to egg

haploid nucleus contains the haploid number of chromosomes so that full complement restored at fertilisation

mid region with mitochondria to provide the energy (from respiration) for movement

acrosome containing enzymes to digest the outer layers of the egg

haploid nucleus – as in sperm

cytoplasm full of energy-rich material

follicle cells

special vesicles (cortical granules) – these contain a substance that helps stop more than one sperm fertilising the egg

zona pellucida (jelly layer) – to stop more than one sperm fertilising the egg

The process of fertilisation

THE ACROSOME REACTION
When the front of a sperm touches the zona pellucida of the egg the acrosome bursts and releases enzymes which digest: a channel in the zona pellucida.

MEMBRANE FUSION
The surface membranes of the sperm and egg fuse allowing the haploid nucleus from the sperm to enter the cytoplasm of the egg.

THE CORTICAL REACTION
Vesicles inside the egg called cortical granules fuse with the cell membrane and release their contents. These cause changes in the surface layer of the egg preventing other sperm from entering.

MEIOSIS IS RESTARTED
The egg is really a secondary oocyte and the presence of sperm causes the 2nd division and meiosis to now occur.

FERTILISATION (the fusion of the nuclei)
Finally, the chromosomes from the haploid sperm and those from haploid egg combine to restore the diploid number; this is fertilisation.

Worked example

Explain how, in mammals, events following the acrosome reaction prevent more than one sperm fertilising an egg.   **(5 marks)**

The sperm cell fuses with an egg cell membrane. Cortical granules then move to the egg cell membrane. Here, they fuse with the membrane, and undergo exocytosis releasing their contents into the zona pellucida; this is called the cortical reaction. This causes hardening of the zona pellucida and formation of fertilisation membrane.

In questions like this it is important to make sure you get the sequence of events correct, think of the flow chart.

Now try this

Describe three structural differences between a human sperm cell and a human egg cell.   **(3 marks)**

Genes and chromosomes

Genes are packaged together on **chromosomes**. A **locus** (plural loci) is the name given to the particular location of a gene on a chromosome. In humans there are 46 chromosomes. When two genes are on the same chromosome they are said to be **linked**, and if it is a sex chromosome, they are **sex-linked**.

Mendel's peas

In the 1850s Gregor Mendel analysed the patterns of inheritance of characteristics in garden peas (for a reminder of **Mendel's Laws** see page 37). He crossed pure-breeding, homozygous, tall purple-flowered plants with short, white-flowered plants:

- He found all the offspring were tall and purple-flowered (called the F1).
- On crossing these further he found tall purple / short purple / tall white / short white in the ratio 9:3:3:1

Mendel concluded that the inheritance of one pair of factors is independent of the inheritance of other pairs.

parent	tall, purple flowers	x	short, white flowers

F₁ generation — tall, purple flowers

F₂ generation — selfed

tall, purple flowers / tall, white flowers / short, purple flowers / short, white flowers

Linkage

It is now known that Mendel's results only happen if genes are on different chromosomes (or far apart on the same chromosome) and **independently assort** (you will learn about independent assortment on page 48). Any two genes with loci on the same chromosome are **linked** and will be passed on as a **pair** to the same gamete.

parent's phenotypes	long wings broad abdomen	vestigal wings narrow abdomen
parent's genotypes	VgVgAA	vgvgaa
gamete genotypes	Vg A	vg a
F₁ generation	Vg A vg a	all long wings broad abdomen

F₂ generation

gametes from mother

gametes from father

	Vg A	vg a
Vg A	Vg Vg A A	Vg vg A a
vg a	Vg vg A a	vg vg a a

In fruit flies, short wings (vestigial, vg) is recessive to long wings (Vg). Narrow abdomen (a) is recessive to broad abdomen (A). When F1 normal flies are crossed with other flies (e.g. VgvgAa × VgvgAa) they do not produce the full range of characteristics expected if the genes were independently inherited. This means the genes are **linked**, so only VgA or vga gametes are produced by each parent.

Sex linkage

Genes with loci on the sex chromosomes (X and Y) are said to be **sex linked**.

Since men have only one X chromosome, if they inherit a faulty gene on the X chromosome they do not have another X chromosome (as women do) with the non-faulty gene. Men suffer from some conditions, such as red-green colour blindness, much more commonly than women because of this.

You are told the probability of them having a colour blind boy is 1:2, so this tells you that the mother must be a carrier; if she were not, the probability of their son being colour blind would be zero.

Now try this

Fruit flies have 8 chromosomes in a body cell. Deduce how many groups of linked genes they have.

(2 marks)

Worked example

A woman has normal vision, but her husband is red-green colour blind. Draw a genetic diagram to show the probability of their son being colour blind as 1:2.   **(2 marks)**

	Xᶜ	Y
Xᴺ	XᶜXᴺ	XᴺY
Xᶜ	XᶜXᴺ	XᶜY

N = normal vision allele
C = colour blind allele

Meiosis

Meiosis produces non-identical gametes (sex cells) that have one set of chromosomes (**haploid**).

Role of meiosis

- to reduce the number of chromosomes by a half (**diploid** cell to **haploid** cell) to avoid a doubling in each generation; this is done by the **reduction division**

- to ensure genetic variation (following sexual reproduction) through the production of non-identical gametes.

Meiosis

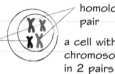

homologous pair

homologous pair

a cell with 4 chromosomes in 2 pairs

note: each chromosome has 2 sister chromatids

meiosis 1 (first division)

the homologous chromosomes pair with each other in the middle of the cell

the homologous chromosomes are pulled apart

the two products of meiosis 1

meiosis 2 (second division)

the two products of meiosis 1 divide again (chromatids separate)

to give 4 cells which become gametes (sex cells)

Each chromosome replicates to create two sister chromatids *before* homologous pairs line up at the start of meiosis I. In order to get the right number of chromosomes in each gamete after meiosis is complete, this second division (meiosis II) is required. This is just like mitosis.

Reduction division

In the reduction division, cells with $2n$ chromosomes give rise to cells with n chromosomes. Using an organism with 8 chromosomes in each cell as an example: if a cell with 8 chromosomes in 4 pairs fused with another similar cell in sexual reproduction, the resulting fertilised egg would have 16 chromosomes, which would not be viable for that organism.

Generation of variation

Meiosis generates variation in two ways:

- **Independent assortment** – the chromosomes can go to either end of the cell when they are being pulled apart, leading to variation by separating alleles into different new cells

- **Crossing-over of alleles** – when the homologous chromosomes are paired at the beginning of **meiosis 1**, a process called **crossing-over** occurs, which also leads to variation by allele separation.

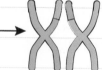

before crossing-over

after crossing-over: genetic material has been swapped

Incomplete linkage

If genes have their loci on the same chromosome (see page 47 for a reminder of sex linkage) but they are far apart, crossing-over can occur between the genes and then they can assort independently.

Worked example

For the cell with four chromosomes shown at the start of the meiosis diagram above, draw out the other possible results of independent assortment at the end of meiosis 1.

(2 marks)

🧮 **Maths skills** To test the fit of a predicted and an actual genetic cross you would choose the Chi squared test.

Now try this

Explain how meiosis gives rise to genetic variation in gametes. **(3 marks)**

The cell cycle

The **cell cycle** is a regulated process in which new cells divide into two identical daughter cells. It consists of three main stages: **interphase**, **mitosis** and **cytokinesis**.

Main events in the cell cycle

interphase:
- the cell grows
- new organelles are made
- towards the end, chromosomes are replicated

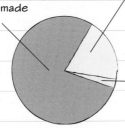

mitosis:
there are 4 stages
- prophase
- metaphase
- anaphase
- telophase

cytokinesis:
the cytoplasm divides to form 2 new cells

Interphase

A very busy time for the cell. It has three stages:

1. G1 (or Gap 1) when growth and protein synthesis occur

2. S (synthesis of DNA) when the events of DNA replication (explained on page 34) occur

3. G2 (or Gap 2) when further growth and protein synthesis occur.

Control of the cell cycle

Control happens at a number of checkpoints between one phase and another (for example, the G1 to S transition). Cyclins build up and attach to enzymes called cyclin-dependent kinases (CDKs). These complexes bring about the next stage in the cell cycle.

In mitosis, the daughter cells are **diploid**.

The stages of mitosis, the process of nuclear division

1. PROPHASE

chromosomes become visible and the membrane of the nucleus breaks down; the nucleolus disappears, the spindle forms and, at the end, centrioles move to either end of the cell

2. METAPHASE

chromosomes, each consisting of two sister chromatids, line up at the equator of the cell; they attach to spindle fibres, which are attached to the centrioles

3. ANAPHASE

spindle fibres contract, pulling the chromatids apart

4. TELOPHASE

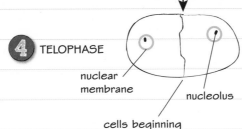

chromosomes decondense and become invisible; membrane of nucleus and nucleolus reform

nuclear membrane

nucleolus

cells beginning to split apart

KEY

chromosome with two separate chromatids

spindle fibre

centriole

Describe the events that occur in mitosis from the start of prophase up to the beginning of anaphase. **(4 marks)**

Chromosomes condense and become visible as two chromatids. The nuclear membrane breaks down and centrioles move to opposite ends of the cell. Spindles arise from centrioles and attach to chromosomes on the equator by centromeres. The spindle fibres start to contract as the cell enters anaphase, pulling the chromatids apart.

Make sure you answer the precise question asked. Here, you must not end until you have said something about anaphase.

How many cells will be present after four divisions of a zygote after fertilisation? **(1 mark)**

Mitosis

Cell division by **mitosis** produces daughter cells that are identical copies of the dividing cell.

Growth

Growth involves an increase in both the number of cells and cell size. Between one mitotic event and another, the cells grow.

In animals, cell growth can occur all over the body and usually stops at maturity.

In plants growth only occurs at special growing points called **meristems**.

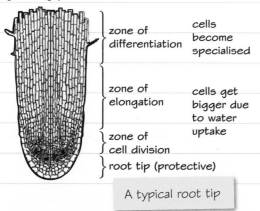

zone of differentiation — cells become specialised

zone of elongation — cells get bigger due to water uptake

zone of cell division

root tip (protective)

A typical root tip

For **core practicals** you need to know **what** you did but also **why** you did it. A table like the one shown would be a useful thing to make for all your core practicals. Safety is vital too.

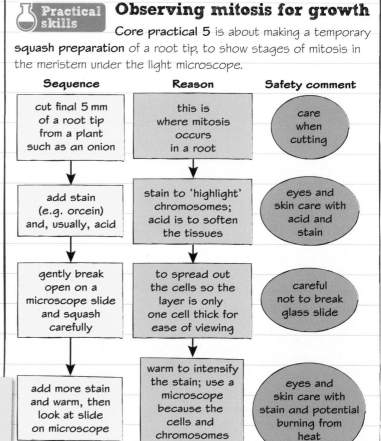

🧪 **Practical skills**

Observing mitosis for growth

Core practical 5 is about making a temporary **squash preparation** of a root tip, to show stages of mitosis in the meristem under the light microscope.

Sequence	Reason	Safety comment
cut final 5 mm of a root tip from a plant such as an onion	this is where mitosis occurs in a root	care when cutting
add stain (e.g. orcein) and, usually, acid	stain to 'highlight' chromosomes; acid is to soften the tissues	eyes and skin care with acid and stain
gently break open on a microscope slide and squash carefully	to spread out the cells so the layer is only one cell thick for ease of viewing	careful not to break glass slide
add more stain and warm, then look at slide on microscope	warm to intensify the stain; use a microscope because the cells and chromosomes are small	eyes and skin care with stain and potential burning from heat

Asexual reproduction

Asexual reproduction takes various forms in different organisms.

- Simple microbes: splitting in two is a simple form of reproduction called **binary fission**
- Other microbes: **budding** occurs in which the two products form, but are not the same size
- Higher organisms: especially plants, asexual reproduction is called **vegetative reproduction**, for example the runners produced by strawberries
- Animals: asexual reproduction is rare in higher animals but is known in lizards; it is more common in simpler animals, for example greenfly reproduce asexually throughout the summer so that their numbers increase rapidly to take advantage of plentiful food supply.

Worked example

Calculate the mitotic index (MI) for the photo shown. **(3 marks)**

$$MI = \frac{\text{number of cells in mitosis}}{\text{total number of cells viewed}} \times 100\%$$

There are 86 cells in the photo.

The number of mitotic cells = 12

$$MI = \frac{12}{86} \times 100$$

$$= 14\%$$

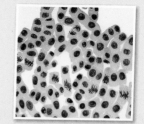

🧮 **Maths skills** It is a good idea to roughly estimate percentage calculations. Here, it is **approximately** 6 out of a 100, which is 6%. So, an answer of 7.14% seems reasonable.

Now try this

Explain why ethanoic orcein and concentrated acid are added to a root tip before it is placed under the microscope to view cells in mitosis.

(2 marks)

Exam skills

This exam-style question uses knowledge and skills you have already revised. Have a look at pages 49 and 50.

Worked example

(a) The photographs show the main stages in the process of mitosis. What is the correct order in which they occur?   **(1 mark)**

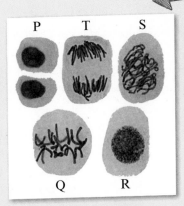

A R → S → T → Q → P

B R → S → Q → T → P

C S → R → T → Q → P

D P → S → Q → T → R

B

> Multiple choice questions are often considered to be easy, but this is not the case. You should approach them by eliminating the three answers you think are wrong. Then, as a double-check, make sure the one that is left is correct. In this case, you can eliminate D, as the end product should have **2** cells (P). The first two steps involve the chromosomes becoming more distinct, so that suggests R and S and they are more distinct in S than R, so that eliminates C. It is now just a question of which way round are T and Q. In Q, the chromosomes are lined up in the centre, in T they are separating, so the answer is **B**.

(b) Explain the procedure carried out to be able to view cells as shown in the photograph. You should start with fresh root tips from an onion.   **(5 marks)**

The root tips are fixed in ethanoic ethanol to preserve the structure of the cells.

A 2–3 mm piece is removed from the tip because this is where dividing cells are located.

This should now be placed in concentrated HCl to macerate the tissue, that is, dissolve the 'glue' between the cells.

The tip is now placed on a slide and ethanoic orcein is added to stain the DNA.

Finally, a cover slip is added and pressed firmly to separate the cells from each other.

> **Practical skills** This is a **core practical** (CP5) and, as such, you should know how it is done. In addition, the question asks you to **explain** the procedure. In explanations of practicals, or ones where you have to say how you would modify a practical, you need to say **why** you did what you did as well as **what** you did.

(c) A student decided to investigate daily rhythms in mitosis in a grass. The following data were obtained. Compare and contrast the pattern shown for shoot tips and root tips.   **(4 marks)**

> Notice the use of the command words **compare** and **contrast**. If you are asked to do this, you must point out **at least** one similarity and one difference. All the points you make must be different from each other.

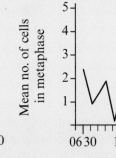
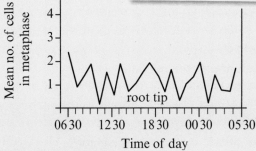

Both show a change in the number of cells undergoing mitosis throughout the day.

Mitotic activity in the shoot tip is much higher than that in the root tip.

Shoot tips show three distinct peaks at 12 30, 18 30 and 00 30, that is, every six hours.

Although the root tip shows change in numbers through the day, it is much less regular than in the shoot.

Stem cells and cell specialisation

Stem cells are undifferentiated (unspecialised) cells that keep dividing and so can give rise to other cell types. If the stem cells can divide to make all other cell types, they are **totipotent**; if they can only make some cell types, they are **pluripotent**. There are two types of stem cell, embryonic and adult stem cells. You need to know about **embryonic cells**.

Embryonic stem cells

At the point when an embryo consists of eight identical cells, all the cells are **totipotent**, which means they can develop into any other body cell. Each cell could become a full human being.

By the time the embryo has become a **blastocyst**, a hollow ball of cells with an inner cell mass of 50 cells, these 50 cells are **pluripotent**. They can become most, but not all, of the 216 human cell types.

Uses of stem cells

Stem cells can form many different kinds of specialised cell, so they could be used to treat some medical conditions that involve cell loss/damage. For example:

- Parkinson's disease – a progressive loss of nerve cells in the brain that are involved in muscle control
- Multiple sclerosis (MS) – the electrical insulating layer surrounding nerve cells breaks down
- Type 1 diabetes – caused when cells in the pancreas produce less than the normal level of insulin in response to a rise in blood glucose concentration
- Burns – skin cells damaged so cannot be replaced.

Ethical issues when using stem cells

- When does an embryo become a human with rights? Should there be a maximum age for embryos used in research? Should human embryos be used at all?
- Is it acceptable to use human embryos specially created for research?
- Is it acceptable to fuse an adult human cell with a human egg cell to create new stem cells?

Therapeutic cloning

A practical problem with transplanting organs is rejection by the immune system. If an organ was grown from a person's own cells this problem would be avoided.

A diploid cell is removed from a patient needing a transplant. This cell's nucleus is fused with an ovum, from which the haploid nucleus has been removed, and a diploid cell results. The process is **somatic cell nuclear transfer**. The stem cells arising from this could then be encouraged to become whatever tissue was needed.

Who decides?

Because of the ethical issues involved, society uses scientific knowledge to make decisions about the use of stem cells in medical therapies.

Decisions are made by:

- people working in the stem-cell field, who understand the issues and what is possible
- everyone else, because they can give a range of alternative points of view.

The final decisions about what can and cannot be researched are made by regulatory authorities such as the **Human Fertilisation and Embryology Authority** in the United Kingdom.

Worked example

Give **two** reasons why some people believe it is a good idea, and **two** reasons why some people believe it is a bad idea, to use spare embryos in research.   **(4 marks)**

The research may lead to treatments for various conditions. Also, the spare embryos would be discarded anyway.

However, embryos are potential humans and have a right to life. Also, women having IVF may be pressured into producing surplus embryos.

Now try this

Human stem cell research involves the use of both totipotent and pluripotent stem cells.
Describe the differences between a totipotent stem cell and a pluripotent stem cell.   **(2 marks)**

Gene expression

All cells in an organism contain the same genetic information and yet there are many different kinds of cell (216 in humans). **Specialisation** is achieved by **differential gene expression** or **gene switching**.

Differential gene expression

Cells become specialised because:

- only some genes are switched on
- the switched on genes produce active mRNA which is translated into proteins within the cell
- the protein produced determine the specific structural and functional features of that cell.

Once cells become specialised they often form clusters with each other to form **tissues**. Recognition proteins on the cell membrane ensure that only like cells join.

Tissues are then joined together to make **organs** and organs to makes **systems**.

The role of active mRNA

Active mRNA is translated into proteins within the cell. These proteins control processes in the cell (usually by being enzymes). Proteins can also form part of cell structure.

The epigenetic factors controlling the activation and deactivation of genes (gene switching)

DNA methylation involves the attaching of a methyl group ($-CH_3$), to a cytosine that is next to a guanine, as an epigenetic marker. The methylation stops the transcription of mRNA by preventing the binding of RNA polymerase. Histones are modified by the addition of acetyl or methyl groups. This causes the DNA to be wrapped up tightly which switches genes off.

acetylation of histone proteins, around which DNA is wrapped, switches genes on by loosening its wrapping with histone

DNA demethylation switches gene on by allowing RNA polymerase to bind

presence of lactose switches gene on in *E. coli*

epigenetic factors

substrate

ON
GENE
OFF

epigenetic factors

repressor molecules switch genes off

by binding

substrate

epigenetic factors

DNA methylation switches gene off by stopping RNA polymerase from binding

absence of lactose switches gene off in *E. coli*

The *lac* operon

lac operon

lactose absent

operator gene

gene for β-galactosidase

DNA

repressor molecule

no mRNA produced

RNA polymerase cannot bind

lactose present

operator gene

DNA

lactose

inactivated repressor unable to bind to operator gene

mRNA coding for β-galactosidase is made

For the *lac* operon, lactose (a disaccharide sugar) stops the repressor molecule (coded for by the regulator gene) from preventing RNA polymerase binding to the operator gene region of DNA. This acts to switch the gene on. This means the cell does not waste resources making an enzyme until it needs to.

Worked example

What does the regulator gene of a bacterial operon do?

(1 mark)

Codes for repressor

Now try this

Human bone marrow contains stem cells that can give rise to various types of blood cell including white blood cells. Explain how a stem cell in the bone marrow can become a differentiated blood cell.

(4 marks)

Nature and nurture

The **phenotype** of a living thing is the outward expression of an interaction between the genetic makeup (the **genotype**, nature) and the environment (nurture). This can be seen in the characteristics or features expressed in the organism.

Simple experiments to separate the role of nature and nurture

In non-humans, simple experiments can be done to separate the role of nature (genes) and nurture (the environment).

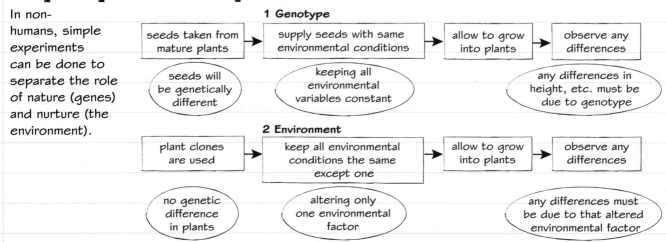

1 Genotype

seeds taken from mature plants → supply seeds with same environmental conditions → allow to grow into plants → observe any differences

seeds will be genetically different

keeping all environmental variables constant

any differences in height, etc. must be due to genotype

2 Environment

plant clones are used → keep all environmental conditions the same except one → allow to grow into plants → observe any differences

no genetic difference in plants

altering only one environmental factor

any differences must be due to that altered environmental factor

An actual study on the subspecies of the herb *Potentilla glandulosa*

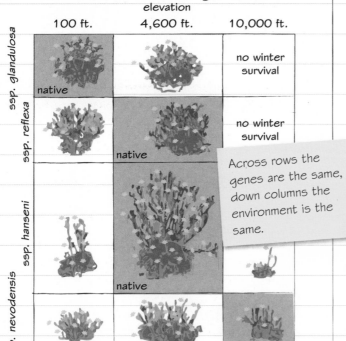

elevation

	100 ft.	4,600 ft.	10,000 ft.
ssp. glandulosa	native		no winter survival
ssp. reflexa		native	no winter survival
ssp. hanseni		native	
ssp. nevodensis			native

Across rows the genes are the same, down columns the environment is the same.

Plants change even though their genes are the same, and this depends on the environment, in this case altitude. The genetic differences determine what the plants look like, their **phenotype**.

Twin studies

Twin studies can be used to separate nature from nurture in humans.

In a pair of identical (**monozygotic**, from one egg) twins reared apart, any differences are assumed to be due to environmental differences as their genes are the same.

Some results:

- eye colour – they have same eye colour, so no environmental effect
- fingerprints – about 95% similar ridge count, so small environmental effect
- depression – about 40% similar, so about half and half nature and nurture.

Worked example

If one identical twin has schizophrenia some research shows there is an 80% chance that the other will also have it. However, if one non-identical twin has schizophrenia there is only a 15% chance that their twin will also have it. Analyse this information to explain the contribution of nature and nurture on the development of schizophrenia.　**(2 marks)**

Genes mainly determine schizophrenia, but there is a small environmental influence.

Now try this

The eye blink reflex is innate. Explain whether nature or nurture is likely to be responsible for the development of an eye blink reflex.　**(2 marks)**

Innate means inborn or inherited, not learned.

Continuous variation

Some characteristics (**phenotypes**) such as human blood group, can only take one form from a restricted number, with no intermediate or overlapping values (making them '**discrete**'). Others, such as human height, can have many different intermediate values (making them **continuous**).

Discontinuous variation

Characteristics like human blood group, eye colour and ear lobe shape, show variation but there are only a few **discrete** states possible. Blood group, for example has four types.

Discontinuous variation like this is often controlled by a gene at a single locus and is often not affected by the environment.

Continuous variation

Characteristics like height and weight show variation with any value across the full range being possible. The graph looks like this:

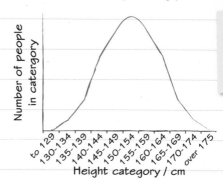

The red curve shows that if enough people were measured the distribution would be continuous.

Polygenic inheritance

Polygenic inheritance is when two or more genes are involved in the inheritance of a characteristic.

At one time, it was thought that continuous variation was due to environment and that genes could only give rise to discontinuous variation, like tall and short. Johannsen showed that a sack of beans with a whole range of weights had, in fact, 19 pure genetic lines of weight, within which there was still variation, this being due to environment. This led to the theory that many genes can produce continuous variation.

Therefore, characteristics with continuous variation are controlled by genes at many loci, i.e. polygenic inheritance.

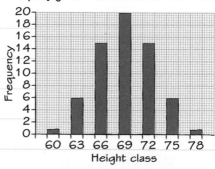

Continuous variation is often displayed by characteristics that are affected by both nature and the environment.

How does polygenic inheritance work?

Imagine plants that are about 60 cm tall.

If they have just dominant alleles of

- gene X they will be 63 cm tall
- gene Y they will be 63 cm tall
- gene Z they will be 63 cm tall.

If they have XYZ they will be 69 cm tall.

A plant with all three genes in double dose (XXYYZZ) will be 78 cm tall.

The recessive alleles x, y and z have no effect on height.

Imagine the cross: Cross   XXYYZZ   ×   xxyyzz
                          (78 cm)   (60 cm)

           F_1 XxYyZz
                          (69 cm)

In the F_2 there will be 27 genotypes ranging from XXYYZZ (78 cm) to xxyyzz (60 cm). For example, XXYyZz would be 72 cm tall, and xxyyZz would be 63 cm tall and so on.

Working through all 27 combinations of alleles gives this distribution

Maths skills   **Deduce** means you need to reach a conclusion from the information provided. The **mode** is the value that occurs the most frequently. The **mean** is the 'arithmetic' average calculated using the formula:

$$\text{mean} = \frac{\text{sum of the amounts}}{\text{total number of amounts}}$$

Worked example

Explain what is meant by the term polygenic inheritance.   **(3 marks)**

There is more than one gene for a single characteristic and these genes are at different loci. Interaction between them gives rise to continuous variation.

Now try this

Deduce the mean and modal heights in the figure above.   **(2 marks)**

Epigenetics

Changes in DNA and the histone proteins in which it is packaged can modify how genes are activated, but how do these changes occur and can they be inherited?

How epigenetic changes are brought about

Review page 53 where you learnt about DNA methylation and histone modification.

With two genetically identical mice, how can one be fat and yellow, and the other slim and brown?

The effect of low levels of methyl groups in the diet are shown in the obese, yellow mice. Chemicals in the diet, such as Bisphenol A (BPA) found in plastic bottles, also stop methylation. Feeding mice a BPA-rich diet plus a supplement with methyl groups gives slim brown mice; giving the same BPA-rich diet but with no methyl supplement gives obese yellow ones. So, environmental factors cause these genetic changes (epigenetic) by switching genes on and off, thus affecting phenotype.

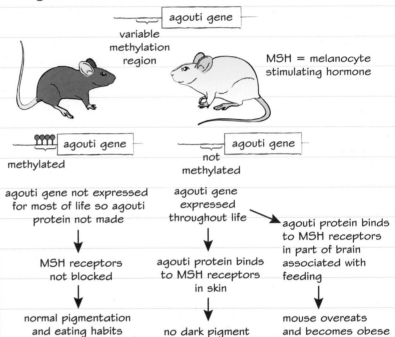

agouti gene

variable methylation region

MSH = melanocyte stimulating hormone

agouti gene — methylated

agouti gene — not methylated

agouti gene not expressed for most of life so agouti protein not made

agouti gene expressed throughout life

agouti protein binds to MSH receptors in part of brain associated with feeding

MSH receptors not blocked

agouti protein binds to MSH receptors in skin

normal pigmentation and eating habits

no dark pigment

mouse overeats and becomes obese

Can epigenetic changes be inherited?

Evidence suggesting that it can includes:

- ☑ 2000 mothers and babies in rural Gambia were studied: mothers who were pregnant in the rainy season had better diets, and their babies had more methylation in six of their genes than those of mothers pregnant in the dry season.

- ☑ People in a Swedish village were studied: it was found that the health conditions the people developed were correlated with the diet of their parents and grandparents.

How strong is the evidence?

The epigenetic memory of cells can sometimes be passed on following cell division. It is also thought it can be transmitted across generations, from parent to child. However, how general such transgenerational effects are in humans and other mammals is a controversial issue. Gametes have changes to their epigenome that have helped determine their specialised structure and function, but the fertilised egg then forms totipotent stem cells, which is probably achieved by the removal of the epigenetic changes. However, it is thought that some epigenetic changes do pass from parent to child. This is a very new field and there is a lot left to discover.

Worked example

Explain why epigenetic changes are unlike mutations of DNA. **(3 marks)**

Epigenetic changes do not alter DNA base sequences but mutations do.

The DNA retains correct information on how to produce a polypeptide with epigenetic changes, whereas mutation may lead to different polypeptides being produced.

Epigenetic changes just alter the degree to which a gene is expressed.

Now try this

Methylated cytosine bases in DNA prevent RNA polymerase from carrying out transcription. Explain why this change in DNA will affect the transcription process. **(2 marks)**

Biodiversity

The vast array of life on Earth, due to millions of years of evolution, is its **biodiversity**.

What is biodiversity?

- Biodiversity comprises every form of life, the genes that give them their specific characteristics and the ecosystems of which they are part. This includes diversity within species, between species and of ecosystems.

- You can talk about the biodiversity of the planet right down to the biodiversity of a piece of woodland.

- Biodiversity tends to be higher where conditions are less harsh, such as in the tropics, and lowest in areas such as the high Arctic where life is on the edge of survivability.

- Biodiversity is threatened as never before due to human activity.

Species richness and species evenness

Species richness measures biodiversity in a particular place. It is expressed as the number of species of a particular group. For example – birds:

- Amazon 1300 species

- UK 574 species

Species richness takes no account of the populations of each species. Two habitats, each with five species of butterfly, may not have the same biodiversity. **Species evenness** is also a factor.

If each of the species has roughly the same population size the habitat is said to have high species evenness. If one species is prevalent and four of the species have low numbers, then the habitat has low species evenness.

A **diversity index** takes into account both the number of species and the size of their populations.

🖩 Maths skills   Diversity index

The index of diversity is used to measure diversity at the **species level**. It can be calculated using the formula:

$$D = \frac{N(N-1)}{\Sigma n(n-1)}$$

Where:

- N = the total number of organisms of all species

- n = the total number of organisms of each individual species

- Σ = sum

- D = index of diversity.

This takes into account species richness and species evenness.

The higher the value of D, the greater the biodiversity. The index allows comparisons of diversity in different habitats.

🖩 Maths skills   Heterozygosity index

This measures the proportion of genes which are present in heterozygous form.

$$\text{heterozygosity index} = \frac{\text{number of heterozygotes}}{\text{number of individuals in the population}}$$

In a population of 27 individuals studied, where
8 were heterozygous the index would be:

$$\frac{8}{27} = 0.30 \text{ (to 2 sf)}$$

Such numbers are very useful for comparisons.

Worked example

The Lundy cabbage flea beetle is known only from the island of Lundy, where it feeds only on the Lundy cabbage. Explain why it is described as endemic. **(2 marks)**

Because species which are found in only one defined region (such as one country or on just one island) are said to be **endemic** to that region.

Adaptation to niches

The **niche** of a species is the way in which the species exploits its environment (for example, a rabbit is a grassland herbivore). They become adapted to their environment by the process of **natural selection**.

Unique niche

Each species has its own niche. If two different species are present in the same niche at the same time, there will be competition and one will out-compete the other so that the better adapted will survive.

Grey squirrels now live in most of the UK because they out-competed the native red squirrels after being introduced from America. But, red squirrels now live in a different niche (upland coniferous forest) because unlike the grey squirrels, they are able to gain adequate nutrition from pine seeds.

Adaptation to niches

There are three main ways species are adapted.

- **Behavioural** – a change in the behaviour of an organism to increase its survival chances, e.g. sheep learn to ignore sounds that are not important to them.
- **Physiological** – these are changes in the internal biochemical functioning of the organism in response to an altered environmental stimulus; e.g. some metabolic reactions become less efficient in cold weather so the organism generates more heat to keep warm.
- **Anatomical** – a physical/structural adaptation (external or internal); e.g. cacti have spines to ward off animals that might eat them.

Natural selection can lead to adaptation and evolution

The evolution of penicillin resistance in a bacterium (opposite) demonstrates the basic steps of evolution by natural selection. In many cases the situation is more subtle, with a reduction in reproductive success rather than a failure to reproduce at all.

> The bacteria **evolve** by becoming **adapted** to the new environment. Note that individuals do not adapt, it is the population that adapts through changes in allele frequency.

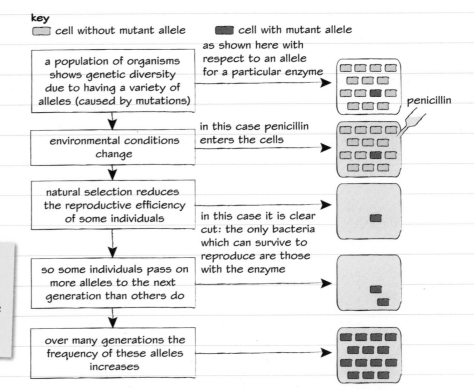

key
☐ cell without mutant allele   ■ cell with mutant allele

a population of organisms shows genetic diversity due to having a variety of alleles (caused by mutations) → as shown here with respect to an allele for a particular enzyme

↓

environmental conditions change → in this case penicillin enters the cells

↓

natural selection reduces the reproductive efficiency of some individuals

↓

in this case it is clear cut: the only bacteria which can survive to reproduce are those with the enzyme

so some individuals pass on more alleles to the next generation than others do

↓

over many generations the frequency of these alleles increases

Worked example

The waxy leaf frog has glands that produce waxy lipids to spread over its skin. This reduces water loss. The waxy leaf frog is active only at night, when it hunts for insects in the trees.

With reference to the waxy leaf frog, explain what the term niche means. **(2 marks)**

The waxy leaf frog eats insects in trees, in hot, dry areas, at night.

Remember that a niche refers to the special behaviour (eats insects at night in trees) an organism exhibits in its environment (hot, dry areas with trees) to survive.

Now try this

*Explain how natural selection might have given rise to the adaptations shown by the waxy leaf frog.

(6 marks)

You need to structure your answer logically, showing how points are related or follow on from each other.

Evolution and speciation

Evolution is a change in the frequency of alleles in a population over time. In order for a new species to form, part of an existing population must become **reproductively isolated** from another part. If the changes in allele frequency are different in the two separated populations, it could lead to new species.

Change in allele frequency

Allele frequency changes because of a change in selection pressure.

For example, in head lice there is an allele that is resistant to the insecticides used in the shampoos designed to kill them. This allele was present before the creation of the shampoos, but it was rare. The use of the shampoos acted as a selection pressure for the resistance allele, which then became more common. A change in allele frequency like this is evolution (but **not** speciation, see below).

The Hardy–Weinberg equation

The **Hardy–Weinberg equation** can be used to detect change in allele frequency in a population.

It shows the relationship between the allele frequencies in a situation where the gene has only two alleles (e.g. T and t). It says that:

$$p^2 + 2pq + q^2 = 1$$

where p = frequency of the dominant allele (T) and q = frequency of the recessive allele (t). Since the gene has only two alleles, $p + q = 1$.

An individual showing the phenotype determined by the recessive allele will have genotype tt.

Individuals showing the phenotype determined by the dominant allele will have genotypes TT and Tt.

When doing Hardy–Weinberg calculations, frequencies should be expressed first as decimal fractions, **not** percentages. The equation predicts that, for any particular values of p and q an unchanging ratio of phenotypes arises; this is the H–W equilibrium.

Do allele frequencies change?

(X) It is very often the case that the equilibrium predicted by Hardy–Weinberg equation does not apply.

(✓) This is evidence for evolution, defined as a change in allele frequency over time in a population.

Worked example

Maths skills Tongue rolling is a characteristic determined by a dominant allele. In a large population, 16% of the people could not roll their tongue. Calculate the expected percentage of homozygous tongue rollers assuming Hardy–Weinberg assumptions apply. **(3 marks)**

Let R be the dominant allele and r be recessive.
Hardy–Weinberg equation is:
$$p^2 + 2pq + q^2 = 1$$
Frequency non-rollers (rr) = q^2 = 16% = 0.16
Therefore, $q = \sqrt{0.16} = 0.4$
Since $p + q = 1$, $p = 1 - 0.4 = 0.6$
Frequency of homozygous roller (RR) = p^2
$p^2 = 0.6^2 = 0.36$
Percentage of expected **homozygous** tongue rollers = 36%

Maths skills When doing Hardy–Weinberg calculations frequencies should be expressed first as decimal fractions, **not** percentages. Note that the frequency of **heterozygous** tongue rollers is: $2pq = 2 \times (0.4 \times 0.6) = 0.48$. To check: $0.16 + 0.48 + 0.36 = 1$

Now try this

Explain what is meant by the term speciation. **(2 marks)**

The classification of living things

Estimates of the number of species of living organisms on Earth puts the total in the millions. In order to study them they must first be **classified**. Living organisms can be organised or 'classified' into major groups, using differences and similarities in phenotypes and genotypes.

Genre of life

Just as when classifying music or films into genre, living organisms can be organised or classified into major groups, using differences and similarities in phenotypes and genotypes.

- The biggest groups of all are the three **domains**, each containing a number of **kingdoms**.

- There are five kingdoms each containing a number of **phyla** (singular **phylum**). Each phylum is further divided into groups which are divided again and again.

The hierarchy of life

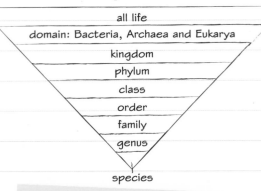

all life

domain: Bacteria, Archaea and Eukarya

kingdom

phylum

class

order

family

genus

species

All living things can be classified into one of three domains.

The definition of species

The idea of the species is a fundamental one.

The most widely accepted definition is:

a group of organisms with similar morphology, physiology and behaviour, which can interbreed to produce fertile offspring, and which are reproductively isolated (in place, time or behaviour) from other species.

The five kingdom system

Animalia – multicellular eukaryotes, all **heterotrophs** (using food made by others)

Plantae – multicellular eukaryotes that are **autotrophs** (they make their own food)

Fungi – multicellular eukaryotes, all heterotrophs

Protoctista – eukaryotes, both autotrophic and heterotrophic forms exist

Prokaryotae – prokaryotic organisms, bacteria and blue-green bacteria

Critical evaluation of new data

The way living things are classified is constantly under review. In the future, the use of DNA analysis of the relationships between species is likely to change things again. This because it will show relationships that have not been apparent by looking at physical similarities.

Worked example

Explain what is mean by the word species. **(2 marks)**

Two organisms are said to be of the same species if they can interbreed to produce fertile offspring.

It is important to be very precise in answers to exam questions. In this case, 'interbreeding to produce offspring' would not be sufficient to gain both marks, neither would 'interbreeding to produce viable offspring'; you need to make it clear that the offspring are fertile.

Now try this

Give **three** ways in which bacteria differ from the eukarya. **(3 marks)**

This question needs you to bring information in from other parts of the course, mainly pages 42–43 (this is a synoptic approach).

The validation of scientific ideas

The scientific community has a critical role to play in the critical evaluation and validation of new evidence. It is illustrated here by evidence for the three domain system of classification and Darwin's theory of evolution.

The case for three domains

> For a long time scientists accepted that life was either prokaryotic (just bacteria and cyanobacteria) or eukaryotic.

> In the 1960s a scientist called Carl Woese began to collect evidence, using **molecular phylogeny**, for another group alongside the bacteria and eukarya, which he called the Archaea. In 1977 he presented his evidence in a scientific journal, although another way for him to have done this would have been through a scientific conference.

Molecular phylogeny is the analysis of genetic material to establish evolutionary relationships between organisms.

> In the following decades other scientists studied the evidence in a process called peer review.
> They would have been looking at his methods, results and conclusions.

The critical evaluation by the scientific community of new data from molecular phylogeny on a new taxonomic group, the Archaea, presented by Woese.

> Towards the end of the 1980s most scientists came to recognise this new group. This recognition came not only from an analysis of Woese's work but also from a repetition and extension of it.

Validating evidence

The scientific process has three key aspects that try to ensure reliability and validity:

• peer review

• dedicated scientific journals

• scientific conferences.

Any research carried out must be published in at least one journal so that it can be read by other scientists. Before the research gets published in a journal, it has to undergo a process called **peer review**. The editor of the journal sends a potential paper to two or three other scientists in the same area of work. They ask:

• Is the paper valid? (Are the conclusions based on good methods and are the data reliable?)

• Is the paper significant? (The paper must make a useful addition to the existing body of scientific knowledge.)

• Is the paper original?

Only if the other scientists agree that the paper is all these things can it be published.

Worked example

Carl Woese's ideas were not accepted at first. Explain how his ideas were critically evaluated.

(3 marks)

His findings were published in scientific journals and delivered at conferences.

Before journal publication the papers containing the ideas would be peer reviewed.

Also other scientists would have repeated his experiments to confirm or validate the findings.

Now try this

Some people still find the theory of evolution controversial. Explain why. **(3 marks)**

Plant cells

Plants are eukaryotes, but their cells differ from those of other eukaryotes such as animals.

The more complex plant cell

Plant cells include all the structures (except centrioles) that are in animals, **as well as**:

- chloroplasts (for photosynthesis)
- a vacuole (to store water and minerals)
- a tonoplast membrane round the vacuole to control movements in and out of it
- a cell wall (for support and protection)
- amyloplasts to store starch
- middle lamella to stick cells together
- plasmodesmata and pits to allow communication between one cell and another.

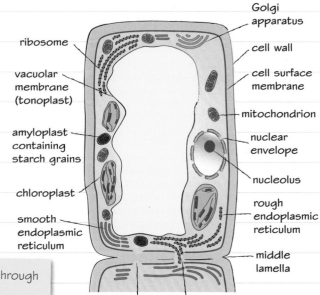

A generalised plant cell as seen through a low power electron microscope

Labels: Golgi apparatus, cell wall, cell surface membrane, mitochondrion, nuclear envelope, nucleolus, rough endoplasmic reticulum, middle lamella, ribosome, vacuolar membrane (tonoplast), amyloplast containing starch grains, chloroplast, smooth endoplasmic reticulum, pit, plasmodesma

EMs of the key organelles

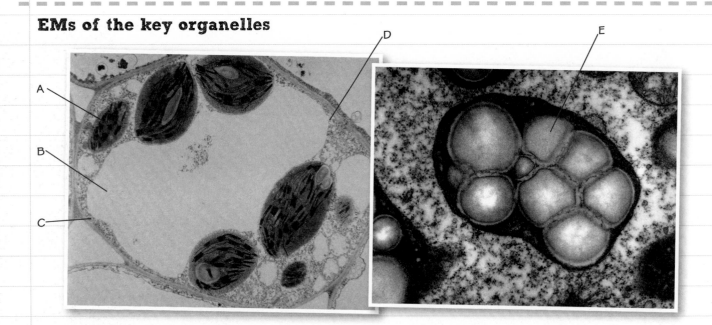

Labels: A, B, C, D, E

Worked example

Identify the structures above labelled A–E.   **(4 marks)**

A: chloroplast; B: vacuole; C: cell wall; D: tonoplast; E: amyloplasts

Now try this

Compare animal, plant and bacterial cells, making reference to chloroplasts, nuclear membrane, cell membrane, ribosomes and centrioles.   **(5 marks)**

A table would be a good way of doing this.

Cellulose and cell walls

Cellulose, found in plant cell walls, is one of the most abundant, and useful, biochemicals on Earth.

Cellobiose

This disaccharide, made from two β-glucose molecules, is only found in cellulose. Read page 10 to remind yourself about glycosodic bonds.

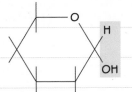

α (alpha) glucose units

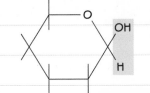

β (beta) glucose units

Each glucose is flipped over with respect to the one it is joined to.

Joining of two β-glucose molecules

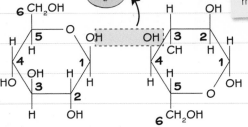

Further cellobiose can join at either end of this molecule to form cellulose chains of infinite length (theoretically).

The structure and function cellulose

In cellulose:

- the chains do not coil but lie next to each other, forming a strong structure
- hydrogen bonds hold the chains together to form cellulose **microfibrils**
- there are parallel straight chains for strong structural function in cell walls.

Compare with starch on page 11.

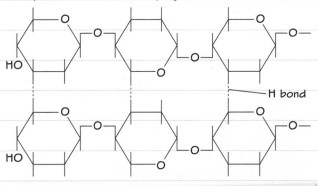

—H bond

Short sections of two cellulose molecules, held together with hydrogen bonds; 60 or so such molecules held together form a cellulose microfibril

The primary cell wall

All plant cells have a primary cell wall made up of cellulose microfibrils. These are arranged in layers, each running at different angles. This leads to strength with flexibility.

On either side of the cell wall

The middle lamella is between cell walls of adjacent cells and contains pectin, which holds cells together. The cell membrane separates intracellular components from the extracellular components (see page 19).

The secondary cell wall

This is laid down in some specialised cells (e.g. xylem and sclerenchyma, see page 64) inside the primary wall. Cells with a secondary cell wall are usually long, and always flexible and strong, which makes them useful to humans as fibres such as flax and cotton.

A cell wall comprises a primary wall and a secondary wall. The primary cell wall has no lignin, whilst the secondary wall is similar but usually thicker and is often impregnated with lignin, which makes it rigid and waterproof.

You should read pages 10 and 11 again before you attempt this question.

Worked example

Compare and contrast the structure of starch and cellulose.   **(4 marks)**

Both are polysaccharides made of monosaccharide monomers, but in cellulose the monomer is β-glucose whereas in starch it is α-glucose. All hydrogen bonding in starch is on one side of the molecule, making it a spiral, but in cellulose they are on both, making it long parallel chains.

Now try this

Explain how the arrangement of cellulose microfibrils contributes to the physical properties of plant fibres.   **(2 marks)**

Transport and support

Plants have three kinds of cell in their **vascular bundles**: **xylem** and **phloem**, which are different transport tissues; and **sclerenchyma**, fibres that support tissue.

Xylem, phloem and sclerenchyma distribution

Vascular bundles are arranged in a concentric circle in most stems, which gives stems both strength and flexibility.

A transverse section of a typical plant stem

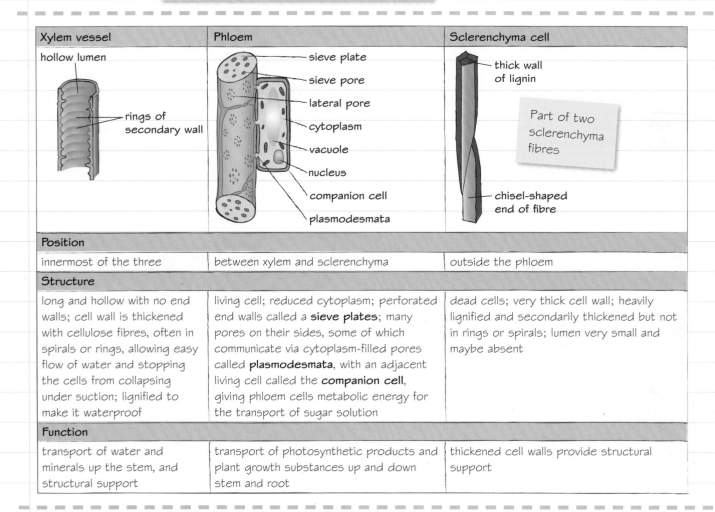

Xylem vessel	Phloem	Sclerenchyma cell
hollow lumen; rings of secondary wall	sieve plate; sieve pore; lateral pore; cytoplasm; vacuole; nucleus; companion cell; plasmodesmata	thick wall of lignin; chisel-shaped end of fibre — Part of two sclerenchyma fibres
Position		
innermost of the three	between xylem and sclerenchyma	outside the phloem
Structure		
long and hollow with no end walls; cell wall is thickened with cellulose fibres, often in spirals or rings, allowing easy flow of water and stopping the cells from collapsing under suction; lignified to make it waterproof	living cell; reduced cytoplasm; perforated end walls called a **sieve plates**; many pores on their sides, some of which communicate via cytoplasm-filled pores called **plasmodesmata**, with an adjacent living cell called the **companion cell**, giving phloem cells metabolic energy for the transport of sugar solution	dead cells; very thick cell wall; heavily lignified and secondarily thickened but not in rings or spirals; lumen very small and maybe absent
Function		
transport of water and minerals up the stem, and structural support	transport of photosynthetic products and plant growth substances up and down stem and root	thickened cell walls provide structural support

Worked example

Label the diagram to show the xylem and the phloem.   **(2 marks)**

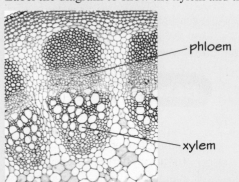

phloem

xylem

Make sure you know where the phloem is: this is often answered incorrectly as the outer bundle of thick walled cells, which are sclerenchyma.

Now try this

Explain why a tree which has its bark removed in a ring will die.

(5 marks)

Looking at plant fibres

 Practical skills You have two core practicals on plant fibres. **Core practical 6** requires you to observe plant fibres and **core practice 8** requires you to find their **tensile strength**.

Core practical 6 – observation practical

Two techniques used:

1 maceration to see separate cells of xylem, phloem and sclerenchyma

You may have used pre-macerated material (e.g. tinned rhubarb). Maceration with acid dissolves the middle lamella between cells, allowing them to be separated.

2 section cutting to see arrangement of cells in stem

You may have used prepared slides rather than cut your own sections. Sections must be very thin to allow light to pass through in a compound microscope.

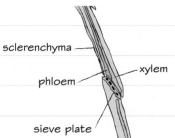

sclerenchyma
xylem
phloem
sieve plate

Core practical 8 – tensile strength

This core practical requires you to find the mass that causes a particular fibre to break.

A typical set of equipment is shown in the diagram. Weights are added until fibre breaks.

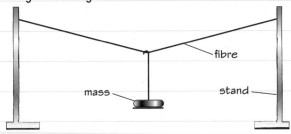

fibre
mass
stand

As always in controlled experiments, non-experimental variables such as temperature, humidity and age of fibre, need to be kept constant.

Exploiting plant fibres

Plant products can be regrown, are not depleting finite stores, and can be recycled or decomposed, so their use is sustainable.

- Molecules of cellulose, arranged as microfibrils, produce a structure that does not stretch but is flexible and has great strength.
- Lignified plant fibres are very resistant to chemical and enzyme breakdown.
- Starch and plant fibres are renewable resources as they come from plants. This means these fibres can be used sustainably.
- Starch can be processed into bioplastic to replace oil-based plastics. Oil is finite so its use is not sustainable.
- They can be burned to release heat energy.

Worked example

Look at the data set.

Repeat	Mass required to break the plant fibre / arbitary units
1	0.206
2	0.203
3	0.207
4	0.205
5	0.209

Calculate the mean and standard deviation (σ) for these data. **(3 marks)**

Mean = sum of all 5 results ÷ 5

= 1.03 ÷ 5 = 0.206

$$SD = \frac{0.00002}{4} = 0.002$$

Maths skills The standard deviation is a measure of how spread out the data are. You can calculate it using the formula:

$\sigma = \sqrt{\dfrac{\Sigma(x - \bar{x})^2}{n - 1}}$ where x is a data value, $\bar{x}$ is the mean of

the data, n is the total number of data in the set and Σ means 'the sum of'. The rules of operations means that you do the calculation inside the root sign first. Of the calculation inside the root sign, you do the bit inside the brackets first.

Note, the answer for the standard deviation (SD) might be displayed on a calculator as, for example, 0.00223607. But it should only be quoted to the correct number of significant figures, which in this case is 4 as that is the number in the original data.

The key word in this question is **valid**.

Now try this

Explain how you would use the apparatus above to enable a valid comparison of the tensile strength of fibres from two different plants.

(5 marks)

Water and minerals

Plants need water from the soil for photosynthesis, transport and cooling. They also need a range of inorganic ions (minerals) to make proteins, nucleic acids, chlorophyll and many other compounds.

Properties of water important to plants

Water is a polar molecule and is a solvent (see page 1); these properties explain most of the useful properties of water.

- Water has a high specific heat capacity; the hydrogen bonds between water molecules keep the temperature in water bodies fairly constant from season to season; this also allows living things to avoid rapid temperature changes.

- Water molecules cohere (stick to each other) and so will stay as continuous columns in the xylem (see page 64); this also means it has high surface tension, which creates a 'skin' that allows organisms, like pond skaters, to move on its surface. Water molecules will also stick to other surfaces, such as glass and xylem tubes; this is **adhesion** and is important in transport in plants (see page 64).

- Water has a maximum density at 4°C; this means that ice floats and insulates the liquid water below, allowing living things to carry on with their life even when temperatures are below freezing.

Minerals

When some substances dissolve in water, they are 'pulled apart' to form **ions**, which can then react. Plants need:

- nitrate ions (NO_3^-) to make amino acids, nucleotides, chlorophyll, ATP and some plant growth substances (e.g. IAA, see page 125)
- calcium ions (Ca^{2+}) used in the formation of calcium pectate in the middle lamella, the glue between plant cells
- magnesium ions (Mg^{2+}) which are an important part of the chlorophyll molecule.

Mineral deficiencies

Plants lacking required minerals will show characteristic symptoms.

- ⊗ Lack of nitrate gives stunted growth with yellowed leaves.

- ⊗ Lack of calcium gives gnarled and misshapen leaves.

- ⊗ Lack of magnesium causes leaves to yellow away from veins but stay green near them.

Worked example

Here are the results of an experiment testing N, Mg and Ca.

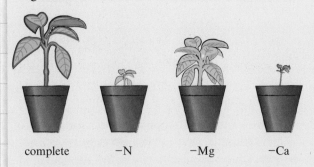

complete   −N   −Mg   −Ca

Explain how you could set up this experiment to give valid results.   **(5 marks)**

The plants are grown from seedlings in either a mineral solution or a soil-less compost watered with a mineral solution. In the case of COMPLETE, the solution contains all the minerals plants are known to need. The −N, −Mg and −Ca solutions contain all the minerals except the one indicated. Non-experimental variables such as temperature, light intensity and humidity should be kept constant. The volume of mineral solution added to each plant should also be controlled. Each treatment should be replicated so that the reliability can be measured. A quantifiable measure of growth such as dry mass should be measured at the end of the experiment.

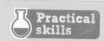

Practical skills   Core practical 7 requires you to investigate plant mineral deficiencies.

Now try this

Explain how the addition of magnesium nitrate might affect the production of oranges.   **(4 marks)**

Developing drugs from plants

From William Withering (1741–1799) to scientists in the present day, people have been looking to plants for useful drugs and testing them to make sure the drugs work and are safe to use.

Clinical trials in the 18th century

Withering, a doctor, learned of a remedy for treating a heart condition. He suspected that the active ingredient, out of the 20 or so ingredients used in the remedy, was the foxglove (*Digitalis purpurea*).

- He made some 'digitalis soup' and gave it to a brewer with swollen limbs and an irregular heartbeat. This man recovered.
- Withering's second 'patient' nearly died, so he decided to stop doing any further trials.
- However, Withering was persuaded to try again. He gave different doses to patients to find the most effective treatment that did not cause severe side effects.
- He carefully recorded all his findings.
- He published his results in a treatise.

The active ingredient in foxglove is digitalin.

Contemporary clinical trials – three-phase testing

A drug can only move on to the next phase when the preceding phase has been passed. It has to pass all three phases before it is released for use.

Carrying out the trials

The trials in each phase would include:

- **a placebo** – the tablet/treatment that appears identical in all ways to the drug except that it is chemically inactive. This is used to see whether the drug has the effect or something else.
- **a double blind trial** – this reduces the chances of bias
- patients are randomly divided into two groups
- one group receives drug/treatment and the other group receives the placebo or standard treatment
- neither the patients nor those recording any changes in the patients know who has received the drug and who has received the placebo or standard treatment.

Historical and contemporary drug trials compared

Similarities:

- ✓ both isolate a possible drug/treatment
- ✓ both initially tested on a small number of patients and then a larger group of patients.

Differences:

- ✗ modern protocols test on animals before phase 1
- ✗ modern protocols have phase 1 testing where the drug is tested on healthy people
- ✗ modern protocols have double blind trials, including using a placebo, undertaken to collect data for statistical analysis.

Worked example

The table shows the actual improvement in patients with schizophrenia at three concentrations of drug treatment. Analyse the information in this table to describe the relationship between the concentration of the drug used and the actual improvement in people with schizophrenia.

(3 marks)

Dose of drug / mg	Actual improvement / arbitrary units
400	6.0
600	12.1
800	12.5

There is an increased improvement with increased concentration. However, this is not linear; the improvement being over 15.25 times larger between 400 and 600 mg of the drug than it is between 600 and 800 mg.

Where you have data and an instruction to **analyse** it to do something, it is a good idea to use the data as here.

Now try this

Give four reasons why a contemporary drug testing protocol is an improvement on the trial used by William Withering.   **(4 marks)**

Investigating antimicrobial properties of plants

Practical skills Plant extracts that have the potential to be antibiotics can be tested using bacteria.

Screening plants for drugs

add nutrient agar with bacteria suspended in it to a Petri dish and allow to cool

↓

add plant material to agar plate; replace lid and seal such that gases can still enter and leave

↓

incubate at 25 °C

↓

observe without removing lid

The bacteria used must not be harmful to humans; cooling allows the agar to set firmly.

Filter paper soaked in plant extract, or plant extract placed in a hole cut in the agar, or plant material laid directly on surface; seal Petri dish and lid so it cannot be opened but air can still get in to stop the development of anaerobic conditions, which encourage harmful bacteria to grow.

Conditions for growth

Bacteria need:

- the correct temperature and pH
- enough nutrients / food (sugars, amino acids)
- enough oxygen
- for toxins not to build up to lethal levels.

In a laboratory culture, one or more of these conditions is not met, so instead of growing to huge numbers, growth levels off after a while.

Aseptic techniques

Aseptic techniques avoid unwanted organisms from being cultured. Culture dishes cannot be sterile as this means nothing is alive within them. Aseptic means that everything is sterile until inoculation, so that **only** those organisms intended to grow are present.

It is important to incubate below a temperature of 37 °C (in schools usually below 30 °C) so that the growth of potential human pathogens is not encouraged.

- ✓ All equipment, including agar, Petri dishes, pipettes and culture tube necks should be **sterile**.
- ✓ Flaming this equipment in a Bunsen flame ensures sterility.
- ✓ **Inoculation** (adding an organism) should be done with a flamed instrument.
- ✓ Lids should be replaced as quickly as possible.
- ✓ Lids should be taped with a cross of sticky tape.

Worked example

In **Core practical 9**, you need to investigate the antimicrobial properties of plants.

An investigation was carried out on the effect of the extraction method on the antibacterial properties of black pepper extract. In one experiment ethanol was used and in another hot water. In both cases the effect on five different bacterial species was investigated.

The data are shown in the table below.

Give evidence from the table to support the hypothesis that (i) ethanol is a more effective extractant than hot water and the hypothesis (ii) that hot water is more effective. **(2 marks)**

Species of bacterium	Mean diameter of zone of inhibition / mm	
	Ethanol extract	Hot water extract
81	27.4	18.2
82	26.2	16.8
83	15.0	29.6
84	25.0	16.4
85	15.0	29.8
mean	21.7	22.2
standard deviation	5.54	6.19

(i) Three of the five species are affected more by the ethanol extract than by the hot water one.

(ii) The mean diameter of the zone of inhibition is greater for hot water than it is for ethanol.

Now try this

(a) Calculate the *t*-value for the data to compare ethanol with hot water, using the formula:

$$t = \frac{|\bar{x}_1 - \bar{x}_2|}{\sqrt{\left(\dfrac{\sigma_1^2}{n_1} - \dfrac{\sigma_2^2}{n_2}\right)}}$$

$\bar{x}$ = mean σ = standard devisaic n = simple size

(3 marks)

(b) Explain why a *t*-test was chosen rather than Chi squared or a correlation coefficient. **(3 marks)**

Conservation: zoos

Due to habitat loss, poaching, pollution and many other reasons, species are becoming endangered in the wild. Zoos have a role in conserving such organisms for the future.

The role of zoos in conserving endangered animals

Zoos have three main roles in conservation.

① **Education** about

- illegal trade in animals and products
- the need for biodiversity (see page 57)
- captive-breeding programmes.

② **Scientific research** on

- control of diseases
- behaviour, to understand needs
- techniques, to improve breeding.

③ **Captive-breeding programmes** to

- increase numbers, reducing the risk of extinction
- release into the wild
- maintain the **genetic diversity**.

The disadvantages of zoos

- animals behave unnaturally
- 95% of species in zoos are not endangered
- animals are shown due to their 'crowd pulling' power
- animals are kept in poor conditions
- exhibition may reduce wild populations.

Why maintain genetic diversity

It is important to maintain genetic diversity through techniques such as captive breeding. However, some zoos may have only a small number of individuals, so inbreeding is likely to occur. This leads to:

- reduced genetic diversity and, therefore, a reduced chance of adapting to environmental change
- increased risk of a genetic condition becoming more common in the breeding population.

Maintaining diversity

Some of the techniques used to reduce inbreeding and maintain genetic diversity:

- do not allow organisms to repeatedly breed with the same partner, possibly by isolating partners
- select partner, possibly by adding a potential partner to a cage, IVF or inter-zoo swapping
- keep a record/database of individuals in captivity and their breeding history, e.g. stud books, so that choice of partners is controlled.

Worked example

The Mount Graham squirrel, *Tamiasciurus hudsonicus grahamensis*, is endemic to Graham Mountain in Arizona, USA. In 2011 its habitat was threatened by wildfires.

The population size of the squirrels is estimated by counting the number of middens where the squirrels bury seeds. The graph below shows the results of surveys carried out from 1991 to 2011.

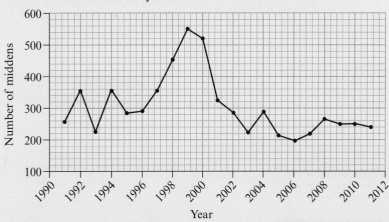

Calculate the percentage change in the number of middens from their highest to their lowest point during this survey. **(3 marks)**

In 1999 there were 550 and in 2006 200. Change is 550 − 200 = 350.

$\frac{350}{550} \times 100 = -63.6\%$. That is a decrease of 63.6%.

You must show the sign if it is a negative change.

Now try this

The black-footed ferret is an endangered species. In 1986, 18 individuals were living in the wild. These were used to start a captive breeding programme. Six zoos are now involved in this programme. Explain how this captive breeding programme in the six zoos ensures that genetic diversity is maintained in this species. **(6 marks)**

69

Conservation: seed banks

Due to habitat loss, climate change, pollution and many other reasons, plant species are becoming endangered in the wild. **Seed banks** have a role in conserving such organisms for the future.

Why conserve wild plants?

Wild plant populations could be wiped out *in situ* by:

- habitat destruction
- climate change
- disasters such as flood and fire
- over-harvesting.

Why worry about a rare plant?

These wild plants may carry genes that can be used in crop plants to confer resistance to pests and diseases.

Seed banks

Seeds from a variety of endangered plants can be stored in a dormant state in seed banks *ex situ*. Seeds rather than living plants are stored because:

- less space is required so more species can be held in the available space
- most plants produce large numbers of seeds so collecting small samples is unlikely to damage the wild population
- easier to store because dormant.

Seed collection and storage

seeds collected from a number of individual plants	Seeds from several plants are likely to have higher genetic diversity than seed from a single plant.
↓	
seeds X-rayed to check for fully formed embryos	This allows only viable seeds (ones that could germinate) to be selected for storage.
↓	
seeds dried to remove water	Reducing water content increase the length of the time a seed can be stored and remain viable, e.g. a 1-2% decrease in seed water content doubles storage time.
↓	
seed stored in the cold, e.g. −20°C	Reducing the temperature also increases the length of time a seed can be stored and remain viable. A 5°C drop in temperature doubles storage time.
↓	
some seeds periodically germinated to check viability	Some seeds are planted to check that they will germinate and grow.

if less than 75% germinate, those that did germinate are allowed to grow into mature plants that produce their own seeds which are then stored	if 75% or more germinate, the remaining seeds are retained in cold storage and can be checked again for viability

Issues with seed banks

- Ⓧ stored specimens have to be replanted as they lose viability
- Ⓧ only some biodiversity can be stored
- Ⓧ certain seeds cannot be stored this way
- Ⓧ seed banks are expensive to build and run; power has to be sustained to keep the seeds very cold.

Worked example

Explain the conditions used for the storage of seeds in seed banks.

(3 marks)

They are kept dry and cold to prevent enzyme activity, the germination of seeds and microbial growth.

Now try this

The only known population of Shiny Nematolepis was destroyed by the bushfires in South East Australia in 2009. However, seeds had already been stored at the Millennium Seed Bank at Kew and were used to restore this plant species and to ensure its survival.

Explain how these seeds would have been selected for storage in the seed bank.

(3 marks)

Exam skills

These exam-style questions uses knowledge and skills you have already revised. Have a look at pages 58, 59 and 60

Worked example

(a) State the correct order of the taxa below
Domain, Kingdom, Species, Class, Phylum, Family, Genus, Order   **(2 marks)**

Domain Kingdom, Phylum, Class, Order, Family, Genus

> Think of a mnemonic to remember the order of the taxa, for example:
> Do Koalas Prefer Chocolate Or Fruit Generally Speaking?

(b) Explain why all species have a binomial name.   **(1 mark)**

All scientists worldwide will call the species by the same name so there will be no confusion as to the species being studied by all of them

> Before binomial classification, species often had long, descriptive names, plus a common name. These names were often different in different countries, leading to a lot of confusion!

(c) State the names of the two species on the phylogenetic tree, which are most closely related.   **(1 mark)**

- Vitis vulpin
- Acer rubrum
- Rubus phoenicolasuis
- Rubus allegheniensis
- Rosa multiflora

> Phylogenetic trees are based on comparisons of DNA, RNA, and amino acid sequence in key proteins, as well as similarities in physical characteristics.

Rubus phoenicolasius and Rubus allegheniensis.

(d) State the features of all species in the kingdom Plantae.   **(2 marks)**

All species in the Plantae kingdom are autotrophic.

Their cells are eukaryotic and contain chloroplasts.

> Autotrophic means that they make organic compounds from inorganic compounds, using light or chemical energy.
> Eukaryotic means that the cells contain a nucleus, as well as other organelles.
> A selective pressure is anything that can affect the survival of an organism.
> For example, competition for resources, predation or disease.

(e) In a shaded environment, taller plants have a selective advantage. Explain how natural selection leads to taller plants.   **(4 marks)**

The amount of sunlight is a selective pressure. Taller plants are more likely to get enough sunlight and survive. Taller plants are more likely to cross-pollinate with other plants. Seeds from taller plants will share the same tall characteristics.

> A selective pressure is anything that can affect the survival of an organism.
> For example, competition for resources, predation or disease

Ecosystem ecology

Organisms are found living together in communities in particular places (habitats) in measurable numbers (populations) within ecosystems. In addition, each species occupies a **niche** which accounts for where it lives and how abundant it is.

Some definitions

- **Ecosystem** – a community of living organisms and their physical environment in an area, which is self-sustaining, e.g. a woodland, a pond
- **Habitat** – the environment where a species or a group of species live
- **Community** – populations of living things interacting with each other in an area
- **Population** – a group of organisms of the same species living together in the same area at the same time.

Biotic and abiotic factors

The distribution and abundance of organisms in their habitat are controlled by:

Biotic (living) factors	Abiotic (non-living) factors
• competition, between individuals (intraspecific) or species (interspecific) • grazing • predation • disease • parasitism, one species benefits, one is harmed • mutualism, both species benefit	• solar energy input • climate • topography (altitude, slope, aspect) • oxygen concentration (especially in water) • edaphic (soil) factors (pH, mineral ion concentration, texture) • pollution • catastrophes, for example earthquakes

Niches

- The **fundamental niche** describes the abiotic factors in which a species could exist.
- The **realised niche** describes where it actually *does* exist; the realised niche takes into account other species.

(You learned about niches on page 58.)

(You learned about niches on page 58.)

Worked example

Two species of flatworm, *Planaria montenegrina* and *P. gonocephala*, were studied.

They were found living in streams both alone and together. When alone they were found in the temperature ranges: *P. montenegrina* 8.5–19 °C and *P. gonocephala* 8.5–23 °C. At temperatures above 23 °C, neither was found.

In a stream where both flatworms occurred together, the following distribution pattern was found.

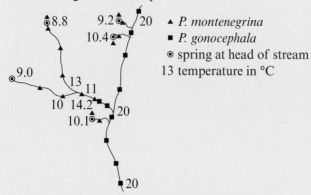

▲ *P. montenegrina*
■ *P. gonocephala*
◉ spring at head of stream
13 temperature in °C

Analyse the information to explain this pattern assuming that both species eat the same type of food.

The fundamental niche of *P. montenegrina* is between 8.5 and 19 °C, that of *P. gonocephala* between 8.5 and 23 °C. The realised niche of *P. gonocephala* when it is found with *P. montenegrina* is much smaller; it does not occur in the colder parts of the stream. The reverse is true of *P. montenegrina*; it is not found in the warmer parts. Thus *P. montenegrina* out-competes *P. gonocephala* at low temperatures and *P. gonocephala* outcompetes *P. montenegrina* at higher temperatures.

Now try this

Explain how the shape of aquatic animals' mouths might be an adaptation to their niche.

(2 marks)

Distribution and abundance

Practical skills   **Core practical 10** requires you to carry out a study on the ecology of a habitat, such as using quadrats and transects to determine the distribution and abundance of organisms, and measuring abiotic factors. There are two main sampling methods: **systematic** and **random** sampling. In both cases a piece of equipment called a **quadrat** can be used to estimate abundance.

Systematic sampling

If there appears to be a change across the area, for example, in an area between the sea and sand dunes, a **transect** is the preferred method.

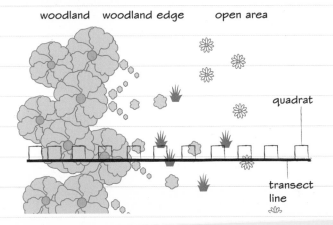

The transect runs from the open area into the woodland. There will be environmental gradients of such things as light, temperature and humidity. These factors can be measured at intervals along the transect.

Random sampling

If two areas appear different and need to be compared, **random samples** could be taken within each area. Quadrats are placed within a grid using random numbers from a set of random number tables.

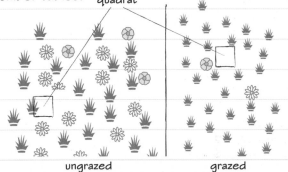

An area grazed by animals is being compared with one that is not. The likely important factor is biotic, that is grazing, but abiotic factors could be measured too.

Worked example

The diagram below shows results of systematic sampling on a shore for two species and the distribution of sandy areas. Analyse the data to explain the pattern of distribution of limpets and *Fucus serratus* between the 5th and 11th metre, and *F. serratus* between the 11th and 18th metre. **(4 marks)**

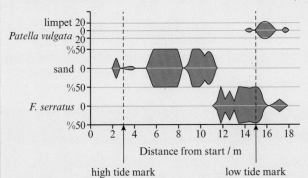

There are no limpets or *Fucus serratus* above 11 metres from the start of the transect, possibly because the beach is sandy there. *Fucus serratus* occurs only from the 11th metre to the 18th metre, possibly because it cannot withstand being out of the water at the top end (11th) and it is eaten by marine animals below the bottom end (18th).

Data from line transects are plotted as a kite diagram. The *x*-axis shows the position on the shore and the *y*-axis the abundance of each species (*F. serratus* and sand as percentage cover and limpets as numbers).

Now try this

Name an abiotic factor that you have measured in a habitat and explain how you measured it. **(3 marks)**

Exam skills

This exam-style question uses knowledge and skills you have already revised. Have a look at page 72 for reminders about sampling in ecology.

Worked example

A study measured the distribution of mussels on a rocky shore near a freshwater stream. The sea rises to cover the stream twice per day. The six numbered sites were sampled.

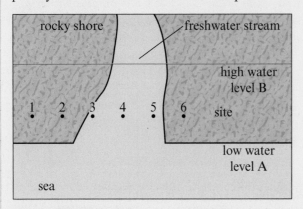

Site	1	2	3	4	5	6
Number of mussels	3	5	15	22	20	2

(a) Discuss the distribution of mussels across the shore.

(5 marks)

Sites 3, 4 and 5 are in the freshwater stream at low water.

Mussels are more abundant in the freshwater region.

This suggests that mussels are more suited to living in fresh water but they can tolerate being covered by seawater for short periods.

Sites 1, 2 and 6 are exposed to air at low tide and it may be that mussels are not able to survive long periods in air.

It is more likely that air rather than salinity is determining distribution, as mussels exist above the low water mark.

Dog whelks prey upon mussels. It takes more than a day for a dog whelk to eat a mussel. The number of dog whelks found at each site was recorded.

Site	1	2	3	4	5	6
Number of dog whelks	7	8	1	0	0	4

(b) Assess the influence of the biotic and abiotic factors on the distribution of the mussels.   **(4 marks)**

Dog whelks are not present in freshwater sites but they are present away from fresh water.

The time taken to eat a mussel means that a dog whelk would be feeding in fresh water at low tide. Mussels will not be eaten where dog whelks cannot feed, which could explain the higher numbers of mussels in the stream. Salinity determines the distribution of dog whelks and predation by dog whelks determines the distribution of mussels.

This question requires you to consider evidence from two different surveys about the effect of biotic and abiotic factors on the distribution of organisms. You should be careful to consider all of the information given. It would be easy to assume that salinity is the only factor but note that the sea will cover all of the sites for significant periods.

Part (a) requires you discuss the information given. You should consider all aspects of the information and use reasoning to explain what you see.

In this answer, two possible factors are considered – the effect of the fresh water and also the time spent out of water at some sites. This is necessary in order to consider all aspects of the data. The answer also includes reasons why salinity is more likely to be the important factor as clearly mussels can survive out of water as they are found there.

The second part of the question adds some additional information and asks for an assessment of the influence of the different factors. The **assess** command word requires that important factors be identified but also that some judgement of importance and a conclusion is reached. Again, information given in the question stem holds the clues as to what is needed.

Statements in the answer include explanations and there is a conclusion covering the whole situation in the final sentence.

Succession

Many ecosystems are unstable and, over time, change to a different one. This process is called **succession**.

Primary succession

Primary succession occurs in an area that is devoid of life.

1 Initial **colonisation** takes place by **pioneer species** (those that can survive conditions where most would die).

2 The environment becomes altered by the pioneer species in ways that makes it unsuitable for them. For example, by allowing the growth of taller plants that will shade them out.

3 This occurs repeatedly, through stages known as seres until a stable **climax community** is reached. This is the final sere. This stage is self-sustaining and stable, usually with one dominant species, and one or a few co-dominant species.

Secondary succession

If the succession starts with soil and some living things are already present, this is called **secondary succession**. It occurs after a disruptive event such as forest fire, hurricane, flooding, or if grazing was stopped in a meadow.

Primary succession on the seashore

Nutrient and moisture-poor sand. Pioneer species with xerophytic adaptations only.

Finally shrubs colonise the area and this allows the development of tall trees. The grazers which might eat the tree seedlings are discouraged by the often spiny shrubs. Woodland is the climax community. In the UK the woodland is usually oak, ash and pine.

The stages (seres) in succession from bare sand to woodland

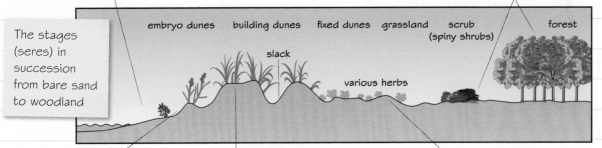

embryo dunes   building dunes   fixed dunes   grassland   scrub (spiny shrubs)   forest

slack

various herbs

Just behind the pioneers comes couch grass which can build dunes by trapping sand. It has long sand-binding roots and grows faster when buried.

Marram grass colonises the next area. This plant can resist water loss by having stomata on the inner surface of its leaves which can roll up in dry weather. It also grows faster when buried.

A wide variety of species now colonise, some of them nitrogen fixers which improve the levels of this important nutrient in the soil. All add more organic matter, improving water holding capacity too.

Worked example

Following the extraction of coal from the ground, the unwanted material was deposited in large heaps. Most of the material is shale fragments made of minerals and clay. The approximate age of a heap (bing) can be estimated by reference to the type of plant community growing on it. This is shown in the table.

Type of plant community	Approximate age of bing / years
lichens and mosses	3–15
grasses and small herbs	15–40
grasses, small herbs and large herbs	40–70
small trees and shrubs	60–80
large trees, small trees and shrubs	80–more than 100

Analyse the information in the table to explain how the type of plant community growing on a bing changes over time.   **(5 marks)**

Lichens and mosses as pioneers form the community. Lichens are able to grow without soil. They break up the rocks forming a shallow soil which allows the growth of mosses and other plants with short roots.

Further changes in soil structure, such as an increase in its ability to retain water and minerals, enable trees and shrubs to grow.

The soil is enriched by the death of plants, which releases organic matter to form humus. The bigger plants eliminate the smaller ones by competition for light, water and minerals.

Now try this

After 100 years the heap will support a stable climax community. Explain why it is stable.   **(3 marks)**

Productivity in an ecosystem

Productivity is the rate of generation of biomass in an ecosystem. It comes from photosynthesis by green plants, and can be measured as a total (gross) amount, or a net amount after the plant has used the portion of productivity it needs for its life processes.

Biomass

Plants make glucose in photosynthesis. This can be turned into other molecules including starch, cellulose, proteins and fats. This **biomass** is food for humans and every other living thing on Earth, including the plants themselves.

Energy flow

The **gross primary productivity** (GPP) is the rate at which energy is made into organic molecules in plants. Plants use some of this energy in respiration (R) (see Topic 7). The **net primary productivity** can be calculated as follows:

NPP = GPP – R

Maths skills GPP, NPP and R are measured in energy units (kilojoules) per square metre per year ($kJ\,m^{-2}\,year^{-1}$).

Finding NPP

The heather plant in a given region is managed by scientists so that big areas of land have plants of one age. 20 areas were sampled, with ages ranging from one year to 20 years old. All the heather was removed from $1\,m^2$ and dried to a constant mass in an oven at $110\,°C$.

The data are shown below.

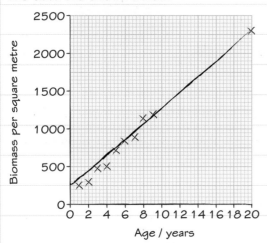

Maths skills The specification asks you to be able to calculate both NPP and efficiency of transfer. From a graph such as this, the **average** NPP is the gradient of the line (you learned about calculating the gradient on page 31).

$$gradient = \frac{change\ in\ y\text{-}values}{change\ in\ x\text{-}values}$$

Maths skills Remember that at least 10% of your marks across all your exam papers will be maths-based.

Worked example

Analyse the data in the heather graph to find the average NPP over the 20-year period. Your answer should be expressed in $kJ\,m^{-2}\,y^{-1}$. Note that one gram of dry heather contains 22.19 kJ. **(4 marks)**

Average NPP is given by the gradient of the line.

That is $\dfrac{2300 - 260}{19} = \dfrac{2040}{19} = 107.3$.

That is in $g\,m^{-2}\,y^{-1}$. Each gram has 22.19 kJ so the NPP as energy rather than biomass is:

$107.3 \times 22.19\,kJ\,m^{-2}\,y^{-1} = 2381.0\,kJ\,m^{-2}\,y^{-1}$

Now try this

Explain why an increase in temperature may cause an increase in NPP. **(2 marks)**

At A level you will need to get used to drawing information from various parts of the course. Here, you need to think about effects of temperature on biological processes.

Energy flow

Energy is transferred between trophic levels in an ecosystem.

Energy transfer between trophic levels

Energy in food is transferred from the primary producers (plants) to the herbivores. They use much of the energy in respiration for movement in the body. Some energy is lost as heat to the environment. The rest is available for other animals or decomposers.

Summary of energy flow in a forest

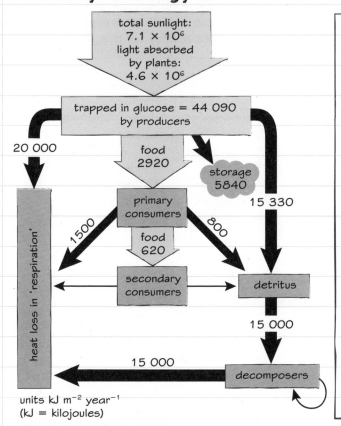

units kJ m⁻² year⁻¹
(kJ = kilojoules)

📟 Maths skills Calculating efficiency

The efficiency of biomass and energy transfers between trophic levels in the diagram opposite is calculated as follows:

1 How much energy is potentially available to primary consumers?

energy trapped (GPP)	−	energy plants use in respiration (R)	=	net primary productivity (NPP)
44 090	−	20 000	=	24 090 kJ m⁻²y⁻¹

2 The transfer efficiency from producers to primary consumers is the amount transferred to the primary consumers' food (in this example, 2920 kJ m⁻² year⁻¹) divided by the amount potentially available to them (24 090 kJ m⁻² year⁻¹).

$$\text{efficiency of transfer} = \frac{2920}{24\,090} \times 100 = 12.12\%$$

Energy transfer efficiencies between trophic levels vary greatly in different ecosystems.

Worked example

The graph below shows the net primary productivity in three environments.

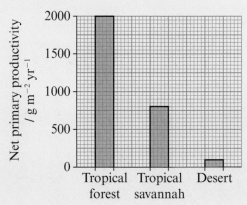

Analyse the data to explain the differences in net primary productivity in these environments. **(4 marks)**

Tropical forest has the highest NPP and desert has the lowest NPP. Tropical forest is 20 times more productive than desert. This is because tropical forest has the best combination of factors affecting NPP. It has plenty of rain to provide water for photolysis in the light-dependent reaction and warm temperatures to maximise enzyme action.

You need to draw from knowledge and understanding in other parts of the course, in this case pages 78 and 86.

Now try this

Look at the data about heather on page 76.

(a) The total incident solar radiation in the Pennines over the heather growth season is 3 144 000 kJ m⁻². Analyse all the information you have about heather NPP in the Pennines to calculate the efficiency of NPP. **(3 marks)**

(b) Explain why this figure is an underestimate. **(2 marks)**

Photosynthesis: an overview

Photosynthesis requires energy from light to split apart the bonds in water molecules. The hydrogen from the water molecules reacts with carbon dioxide to form glucose (a fuel). Oxygen is released into the atmosphere.

Photosynthesis – a summary diagram

light energy 〰〰〰→ $2H_2O$ from the soil

O_2 $4H$ ⟶ $2CO_2$ from the atmosphere

$2(CH_2O)$

The splitting of water by light is called photolysis. A pigment molecule called chlorophyll traps the light energy for this step.

A two-step process

The overall process of photosynthesis is achieved in two linked stages:

1 the **light-dependent reactions** (in which water is split and ATP and reduced NADP are made)

2 the **light-independent reactions** (in which the energy from ATP and reducing power from reduced NADP are used to make sugar).

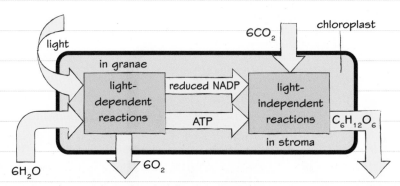

ATP, the energy currency of the cell

Adenosine triphosphate (ATP) provides energy for chemical reactions in the cell.

- When energy is needed, phosphate is removed from the ATP by hydrolysis to give ADP and a phosphate. The energy is released when the phosphate forms bonds with water.

- In the light-dependent reactions, ATP is made using energy from light and is used as a source of energy in the light-independent reactions.

- ATP is also used widely in organisms as a way of transferring energy. It is an intermediate between energy-producing reactions and those that need energy.

ATP ⟶ ADP + Pi + energy

Worked example

The diagram shows the results of an experiment to test the hypothesis that the oxygen given off in photosynthesis comes from water. Air containing $^{18}O_2$ (heavy oxygen) and water containing $^{16}O_2$ (normal oxygen) was bubbled through a suspension of algae.

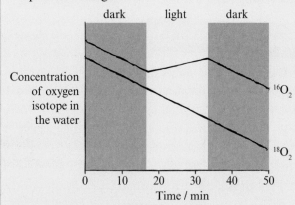

Analyse the data to explain how they support the hypothesis.

The alga produces oxygen from the water it uses in photosynthesis, but only in the light. At all other times both types of oxygen, which are chemically indistinguishable, are being used in respiration, therefore levels are falling due to this, both in the light and the dark. The fall in $^{16}O_2$ in the light is offset by its release from water.

Now try this

The diagram summarises the interconversion of ATP and ADP. Name reactions S and T and substance W.

(3 marks)

ATP $\underset{\text{reaction T}}{\overset{\text{reaction S}}{\rightleftarrows}}$ ADP + substance **W** + energy

The light-dependent reactions

The light-dependent reactions of photosynthesis include the trapping of light energy and the generation of ATP, NADPH and oxygen through the photolysis of water.

The Z-scheme of the photosynthesis light-dependent reactions

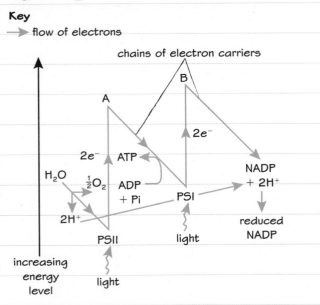

Key
→ flow of electrons

Photosystems PSI and PSII are two special chlorophyll molecules, which can release their electrons when struck by light.

The Z-scheme described

1. A pair of electrons from chlorophyll (PSII) are boosted to a higher energy level by light energy; they are accepted by an electron acceptor (A) and then passed along a chain of carriers. (Electron holes in PSII are filled by electrons from water, split by photolysis; the water is split into H^+ and OH^- and the OH^- interact to give oxygen and new water.)

2. Energy released is used to convert ADP and inorganic phosphate (Pi) into ATP. This process is called **photophosphorylation** (the light-driven addition of phosphate).

3. The electrons then enter another chlorophyll molecule (PSI) where they are again boosted to an acceptor (B) by light.

4. The electrons eventually pass to NADP, with hydrogen from water, to form reduced NADP (NADPH).

Worked example

Explain how oxygen is produced in chloroplasts during photosynthesis. **(3 marks)**

The splitting (photolysis) of water uses light energy. Hydroxyl ions interact to release oxygen and generate new water. Water, specifically OH^-, is the source of the oxygen that plants release.

Now try this

The diagram summarises the light-dependent reactions. Name A, B and C. **(3 marks)**

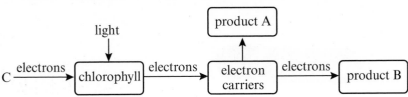

The light-independent reactions

The light-independent reactions (sometimes called the **Calvin cycle**) involve reduction of carbon dioxide using the products of the light-dependent reactions to make simple sugars.

The link between light-dependent and light-independent reactions

The light-dependent reactions make ATP and reduced NADP. These are used in the light-independent reactions (Calvin cycle):

- reduced NADP provides reducing power (electrons)
- ATP provides the energy for the process of making carbon dioxide into carbohydrate.

The fate of sugars made in photosynthesis

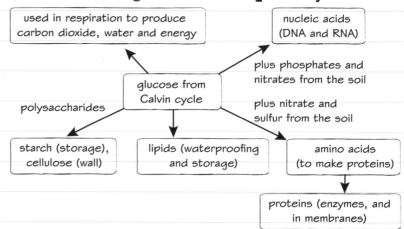

used in respiration to produce carbon dioxide, water and energy

nucleic acids (DNA and RNA)

plus phosphates and nitrates from the soil

glucose from Calvin cycle

polysaccharides

plus nitrate and sulfur from the soil

starch (storage), cellulose (wall)

lipids (waterproofing and storage)

amino acids (to make proteins)

proteins (enzymes, and in membranes)

The Calvin cycle

1 Carbon dioxide combines with a 5-carbon compound called **ribulose bisphosphate (RuBP)**. This reaction is catalysed by the enzyme **ribulose biphosphate carboxylase (RuBISCO)**, the most abundant enzyme in the world.

5 Ten out of every 12 GALPs are involved in the recreation of RuBP. The ten GALP molecules rearrange to form six 5-carbon compounds; then phosphorylation using ATP from the light-dependent reactions forms RuBP.

2 The 6-carbon compound formed is unstable and immediately breaks down into two 3-carbon molecules, **glycerate 3-phosphate (GP)**.

3 This 3-carbon compound is reduced to form a 3-carbon sugar phosphate called **glyceraldehyde 3-phosphate (GALP)**. The hydrogen for the reduction comes from the reduced NADP from the light-dependent reactions. ATP from the light-dependent reactions provides the energy required for the reaction.

$6CO_2$

$6RuBP$ (5C)

$12GP$ (3C)

12 reduced NADP

$12ATP$

$12NADP$

$12ADP + 12Pi$

6ADP

(10GALP)

$12GALP$ (3C)

6ATP

(2GALP)

glucose (6C) (hexose)

☐ from LDRs

4 Two out of every 12 GALPs formed are involved in the creation of a 6-carbon sugar (hexose) which can be converted to other organic compounds, for example amino acids or lipids.

Worked example

Explain how GALP is used to synthesise the cellulose in plant cell walls.   **(4 marks)**

GALP is converted to β-glucose. Glycosidic bonds between C1 and C4 in a condensation reaction join these. This forms the straight chains of cellulose, a polysaccharide of glucose.

Glucose from the Calvin cycle is also converted into nucleic acids, amino acids (then proteins), lipids, starch and cellulose.

Now try this

Which of the following catalyses carbon fixation?   **(1 mark)**

A GALP   B GP   C NADP   D RuBISCO

Exam skills

This exam-style question uses knowledge and skills you have already revised. Have a look at pages 77 and 78 for reminders about photosynthesis and the light-dependent reactions.

Worked example

An outline of some stages of photosynthesis is shown in the diagram. FBPase is one of the enzymes needed for the regeneration of RuBP from GALP.

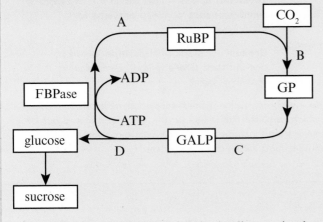

(a) Which of the following points in the diagram is where the enzyme RuBISCO is required?

A   **B**   **C**   **D**   **(1 mark)**

B

(b) Some plants have been genetically modified to reduce the activity of FBPase. The effect of this modification on the rate of sucrose production was measured. The graph shows the results.

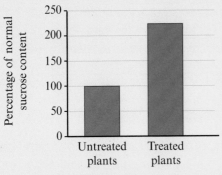

*Analyse these results to discuss the possible benefit of these modified plants in agriculture. **(6 marks)**

Reducing FBPase activity increases sucrose content which will increase yields of crops. If rate of FBPase is lower, more GALP will be available to enzymes forming glucose so more sucrose will be made. However, this will mean less regenerated RuBP will be available, slowing the cycle.

For increased sucrose production there will need to be an alternative source of RuBP. Benefits will only be evident where energy (light) is not limiting as energy will be needed to produce more RuBP.

This is an example of how questions are structured to increase in difficulty. Early parts of a question may be straightforward and the last part more difficult.

The multiple choice question is recall, but in the context of an unfamiliar diagram, so your knowledge needs to be secure. **In the exam**, if you change your mind for an MCQ answer, follow the instructions carefully about how to show this. If you change your mind more than once, show very clearly your final choice.

This part is the challenging question, with **analyse** as the command word. This requires that you identify issues and explore them, giving reasoned explanations. **In the exam**, you may not be directed to use all of the information in the question and data but you should wherever possible. The * here indicates that marks will be rewarded for your ability to structure your answer logically.

The first point goes beyond a simple description by including a relevant consequence. The second and third points make use the first diagram to work out how the chemical pathways are affected. The last two points show what is meant by exploring possibilities in a discussion. Clearly, if more sugar is being made, something else is not, so there must be further issues.

Make sure your answer has a clear structure, and aim to include scientific terminology and biological evidence where relevant.

Chloroplast

 Practical skills The chloroplast is an organelle unique to plants. It is highly adapted to its role in photosynthesis.

The chloroplast – site of light-dependent and light-independent reactions

Thylakoid membrane – a system of interconnected flattened, fluid-filled sacs, where the light-dependent reactions take place.

DNA loop – chloroplasts contain genes for some of their proteins.

Stroma – the fluid surrounding the thylakoid membranes. Contains all the enzymes needed to carry out the light-independent reactions of photosynthesis.

Starch grain – stores the product of photosynthesis.

Smooth outer membrane – which is freely permeable to molecules such as CO_2 and H_2O.

Thylakoid space – fluid within the thylakoid membrane sacs contains enzymes for photolysis.

Granum – a stack of thylakoids joined to one another. Grana (plural) resemble stacks of coins.

Smooth inner membrane – which contains many transporter molecules. These are membrane proteins which regulate the passage of substances in and out of the chloroplast. These substances include sugars and proteins synthesised in the cytoplasm of the cell but used within the chloroplast.

The Hill reaction – Core practical 11

Core practical 11 requires you to investigate photosynthesis using isolated chloroplasts (the Hill reaction). In this experiment, chloroplasts are isolated from leaf cells and illuminated to emit electrons, which are accepted by a blue dye, DCPIP, which goes colourless when reduced.

The procedure also involves the setting up of controls to eliminate possibilities such as DCPIP changing colour due to standing at room temperature in the light or in the dark.

Tube	Leaf extract / cm^3	Supernatant / cm^3	Isolation medium / cm^3	Distilled water / cm^3	DCPIP / cm^3	Purpose	Result
1	0.5	–	–	–	5	experimental tube	colourless
2	–	–	0.5	–	5	control to show DCPIP stays blue in the light	blue
3	0.5	–	–	–	5	control to show no effect in the dark	blue
4	0.5	–	–	5	–	control to show no change in leaf extract	blue
5	–	0.5	–	–	5	control to show no chloroplasts in supernatant	blue

Table showing the five tubes set up for the experiment, including the purpose of each tube and the expected results. Note tube 3 is kept in the dark.

Complete the diagram, which shows the double membrane of a chloroplast, to show the structures involved in the light-dependent reactions of photosynthesis. **(2 marks)**

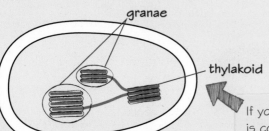

granae

thylakoid

In photosynthesis, in which of the following is light directly involved? **(1 mark)**

A reduction of CO_2 to GP

B synthesis of starch

C release of H^+ from water

D release of oxygen from carbon dioxide.

If you just draw all the structures in a chloroplast, even if it is correct, you will not be answering the question, which asks only for the parts involved in the light-dependent reactions.

Climate change

Scientists believe the climate is changing. Evidence comes from studies of carbon dioxide levels, temperature records, pollen in peat bogs and dendrochronology (tree-ring dating).

The key questions to be asked

- Are CO_2 levels rising?
- Is there global warming?
- Does one cause the other?
- How bad will it get?
- Can humans do anything to combat it?

For more about how bad climate change will get and if humans can do anything to combat it, see page 86.

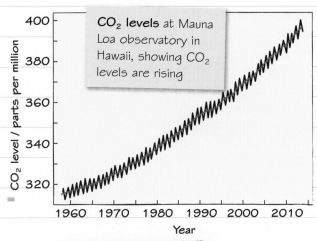

CO₂ levels at Mauna Loa observatory in Hawaii, showing CO_2 levels are rising

Global warming

Long-term data sets allow changes in temperature to be analysed. For example, the Central England Temperature series (CET) has records from 1659 to the present, the longest set available in the world.

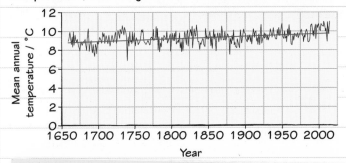

CET data: the data were downloaded from the Hadley Centre and plotted using Excel, which was also used to fit a trend line. Do you think it shows warming or not?

Other evidence

- **Study of tree rings (dendrochronology)** – if the climate is warmer and wetter then tree rings are wider; you can look at tree ring widths over 3000 years into the past and tell a lot about the climate from them.

- **Pollen analysis** – pollen grains are preserved in peat bogs; analysis of the pollen can tell you which plants were growing and so what the climate was like when the peat was formed; deeper in the peat = older.

- **Ice cores** – air trapped in ice when it was formed thousands of years ago can be analysed to give information about temperatures and CO_2 levels in the past.

Worked example

From studies of pollen, Antarctic ice and tree rings, data have been gathered about temperatures and CO_2 levels for thousands of years in the past. These data are shown in the graph.

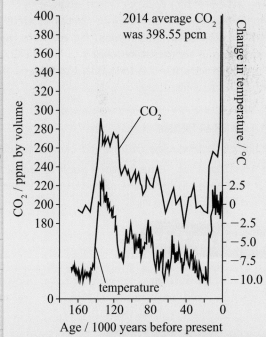

Assess if there is enough evidence here to state that changes in the level of CO_2 cause a rise in global temperature. **(4 marks)**

There is clearly a correlation between the two variables. However, a correlation is not necessarily causal. It could be argued that the temperature rise is causing the rise in CO_2 rather than the other way round. Or that another factor was causing both.

Now try this

Look at the trend line in the Mauna Loa graph. Calculate the percentage increase in carbon dioxide level between 1960 and 2010. **(2 marks)**

Anthropogenic climate change

The causes of anthropogenic (made by humans) climate change include the enhanced greenhouse effect.

Greenhouse Earth

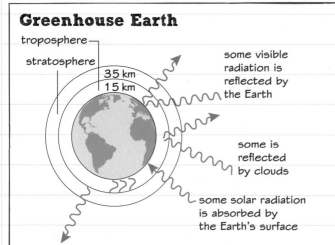

troposphere
stratosphere
35 km
15 km

some visible radiation is reflected by the Earth

some is reflected by clouds

solar radiation

some solar radiation is absorbed by the Earth's surface

some infrared emitted by the Earth's surface escapes and cools down the Earth

some infrared is absorbed by greenhouse gases, warming the troposhere

infrared radiation from the Earth

Greenhouse gases allow radiation to reach the Earth from the Sun. Some of this energy is trapped and the Earth warms up. This is the **greenhouse effect**.

These gases are increasing in the atmosphere due to:

(for CO_2)
- burning fossil fuels
- deforestation, followed by burning

(for CH4)
- cattle produce it
- it comes from rice paddies.

Inputs and outputs of energy to and from the Earth's atmosphere

Predicting the future

Any attempt to predict climate change in the future must rely on very complex computer models to extrapolate data. This means, for example, applying increases in temperature over the last 100 years to projections for the following 100 years. These models are limited by:

- lack of computing power
- lack of sufficient data
- lack of knowledge of how the climate functions
- the fact that some factors such as carbon dioxide emissions or changes in ice cover are very hard to predict.

Predictions for temperatures in New Zealand over the next 100 years

—— assumes no control of greenhouse gas emissions
---- assumes control of greenhouse gas emissions

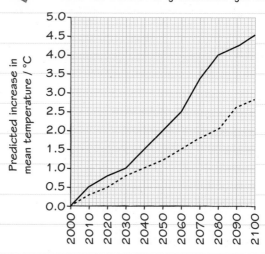

Now try this

Explain the difference between the terms 'global warming' and 'the greenhouse effect'. **(3 marks)**

Worked example

Which one of the following statements most accurately describes the greenhouse effect? **(1 mark)**

A An alternative term for climate change.

B The only cause of climate change during the past 100 years.

C A process which causes the Earth to be warm enough to support life.

D A process which makes it easier for infrared radiation to leave the Earth's atmosphere than for it to enter.

C

People often confuse the greenhouse effect, global warming and climate change. The Earth would be far too cold without its natural greenhouse gases. The effect seen in the last 100 years is better referred to as **the enhanced greenhouse effect.**
A is wrong because the greenhouse effect and climate change are **not** the same. B we do not know this is the case, and D is the wrong way round.

The impact of climate change

Climate change (changing rainfall patterns and changes in seasonal cycles) may affect the distribution, development and life cycles of plants and animals.

The impact of global warming

Global warming

Rising temperatures, changing rainfall patterns, changes to seasonal cycles

Changes to species distribution
Species in the south may move north and vice versa. Alien species may move in and out-compete native ones.

Changes to development
Warming could cause changes in sex ratio in species where sex depends on temperature. Development may be accelerated in some species.

Changes to life cycles
Life cycles of insects and the plants they feed on may become out of synchrony.

The climatic effects of anthropogenic global warming are likely to be complex and very difficult to predict. They will surely be seen in rainfall, temperatures and seasonal cycles.

Temperature

The most obvious climatic change is to a warmer world. Temperature affects the rate of enzyme activity, and because of this it affects whole organisms – plants, animals and microorganisms. They may grow or develop faster; they may die if their temperatures get too high because their enzymes will denature.

The effect of temperature on the initial rate of an enzyme-catalysed reaction. All enzymes have an optimum temperature but it is not always the temperature shown here.

Temperature and enzymes

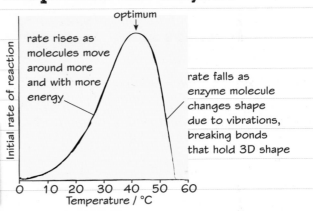

rate rises as molecules move around more and with more energy

optimum

rate falls as enzyme molecule changes shape due to vibrations, breaking bonds that hold 3D shape

 Core practical 12

In core practical 12 you investigated the effect of temperature on the initial rate of an enzyme catalysed reaction including Q_{10}. You learned about measuring the initial rate of a reaction on page 31. The same technique is applied here but substrate concentration, enzyme concentration and pH are kept constant whilst temperature is varied.

$$Q_{10} = \frac{\text{initial rate at } (T + 10\,°C)}{\text{initial rate at } T\,°C}$$

Worked example

Tuataras are reptiles found in New Zealand. The sex of a tuatara, which hatches from an egg, is determined by the nest temperature. At 22 °C or above, a male tuatara will hatch. Female tuataras hatch from eggs incubated below 22 °C. In 2000, the temperature of the nests ranged between 18 °C and 24 °C. Explain how the changes in the mean temperature, shown on page 83, might affect the tuataras.   **(4 marks)**

In 2000, both males and females would hatch. But as temperature rises, more males and fewer females will hatch. This means reproduction rate falls so there will be a fall in population. If nest temperature rises above 22 °C, only males will hatch. However, the lower estimate never reaches this point.

Now try this

Adult tuataras grow to approximately 65 cm in length and feed on small mammals, bird chicks and invertebrates such as insects and worms. Explain how other animal populations on these islands might be affected by changes in the tuatara population.   **(2 marks)**

The effect of temperature on living things

 Practical skills In **Core practical 13**, you will investigate the effects of temperature on the development of organisms. This requires you to understand the effect of temperature on the rate of enzyme activity and its impact on plants, animals and microorganisms.

Hatching rates in brine shrimp

In this practical:

- temperature is the independent variable
- the number of shrimps hatched is the dependent variable
- other variables that might affect the hatching rate are salinity, pH and light level.

It is also possible to study the effect of temperature on seedling growth rates. A fast growing species like radish is ideal for this.

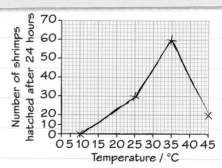

Typical data from a brine shrimp hatching experiment

Field observations and field experiments

There is a clear correlation between higher temperatures and emergence times, the latter being earlier when the former are higher.

1. Insect emergence times

2. Limestone grassland plots were subjected to artificial warming. The three treatments were:
   H = soil temperature 3 °C above surroundings during the winter;
   D = drought, soil received no water in July and August; W = water, additional water added June to September.

There is a clear correlation between higher temperatures and emergence times; the latter being earlier when the former are higher.

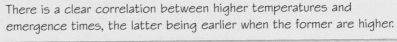

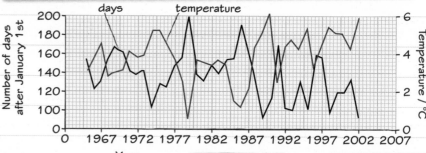

Worked example

With reference to the limestone grassland graph above, explain the likely effect on the abundance of salad burnet in this habitat if the climate became hotter and wetter. **(2 marks)**

It may decrease because the heat has a bigger effect downward than the water has upwards; 10% up with water but 40% down with heat.

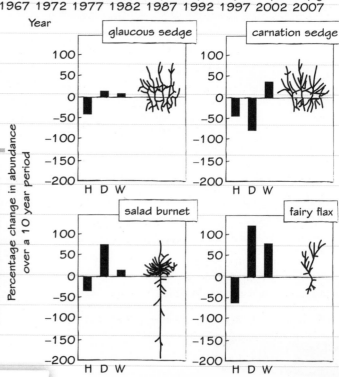

Now try this

With reference to the limestone grassland graph above, if it became hotter and drier, which species would you expect to become less abundant in this habitat? **(1 mark)**

These graphs show the percentage change in abundance of four plant species on limestone grassland, subjected to the three treatments.

Decisions on climate change

Scientific conclusions about controversial issues, like climate change, can sometimes depend on who is reaching the conclusions.

Decisions about climate change

There is little doubt that global warming is happening, but there are still big questions over what is causing it and what we should do about it.

It is quite normal for scientists to disagree but this topic is also a matter for public debate. Non-scientists may not understand the uncertainty and want a clear answer.

The people who will give them this are often not the scientists, but politicians, economists and other policy makers.

Science, politics and economics

The conclusions people reach are often influenced by who funded the research they are doing, and pressures of economics and politics:

- the debate becomes political and then the impassionate methodology of science becomes sidelined
- data may be interpreted with various hidden agendas and this becomes the news rather than the science
- scientists can be accused of being influenced by companies that provide funding.

What can help us decide?

It is clear that there is extra CO_2 being added to the atmosphere by human activity; the cycle shows there are a number of ways we could intervene to offset this.

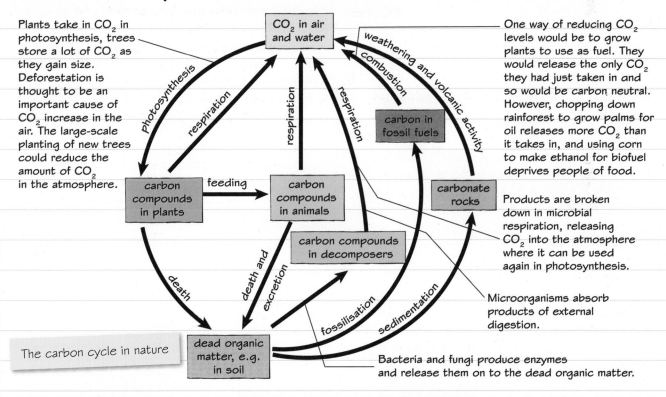

Plants take in CO_2 in photosynthesis, trees store a lot of CO_2 as they gain size. Deforestation is thought to be an important cause of CO_2 increase in the air. The large-scale planting of new trees could reduce the amount of CO_2 in the atmosphere.

One way of reducing CO_2 levels would be to grow plants to use as fuel. They would release the only CO_2 they had just taken in and so would be carbon neutral. However, chopping down rainforest to grow palms for oil releases more CO_2 than it takes in, and using corn to make ethanol for biofuel deprives people of food.

Products are broken down in microbial respiration, releasing CO_2 into the atmosphere where it can be used again in photosynthesis.

Microorganisms absorb products of external digestion.

The carbon cycle in nature

Bacteria and fungi produce enzymes and release them on to the dead organic matter.

Worked example

Explain why the use of biofuels may help to reduce global warming. **(3 marks)**

Biofuels are carbon neutral because they are made from plants, which take up carbon dioxide in photosynthesis.

Now try this

Explain **two** disadvantages of using biofuels to reduce global warming. **(4 marks)**

Evolution by natural selection

Evolution is a change in allele frequency. It can come about through gene mutation and natural selection. The fact of evolution is still disputed by some but is further supported by new evidence.

Hardy–Weinberg

On page 59 you learned about the Hardy–Weinberg equilibrium. It was mentioned that this theoretical state of affairs is rarely met in nature.

For a Hardy–Weinberg equilibrium to apply:

- mating must be random
- the population must be large
- there should be no movement of organisms into or out of the population (no migration)
- there must be no mutations
- there should be no selection pressure, that is, nothing that favours one allele over another.

If *any* of these conditions do not apply, the population will not be in equilibrium and allele frequencies will change.

Selection and mutation

A departure from any of the five conditions will lead to a change in allele frequency.

Although each condition can lead to allele frequency change, it will be directionless. Selection acting through environmental pressure can be directional and lead to adaptation and maybe, eventually, speciation (see page 88 for more on speciation).

New evidence for evolution

New scientific methods such as molecular phylogeny (see page 61) have enabled scientists to identify new types of evidence to support Darwin's theory of evolution.

- The DNA molecule is the same in all organisms. This supports Darwin's idea of descent from a common ancestor.
- DNA and proteins contain a record of genetic changes that have occurred over time. By studying DNA (**genomics**) and proteins (**proteomics**) these changes can be identified.
- Assessing the speed of mutation in DNA has shown that species have evolved over vast periods of time, as Darwin thought. Look back at page 61 to see how the scientific community validates new evidence.

Worked example

The table shows the number of differences in the protein cytochrome c between species. Use it to draw a phylogeny diagram to show degree of relatedness for these species and groups.   **(4 marks)**

	horse	zebra	rabbit	chicken	duck	snake	moth	rice
horse		1	6	11	10	21	29	47
zebra			5	10	9	19	28	46
rabbit				8	6	16	24	46
chicken					3	18	29	48
duck						17	27	47
snake							29	46
moth								45
rice								

rice   moth   snake   chicken   duck   rabbit   zebra   horse

5

10

20

30

40

The more distantly related two groups are, the more differences there will be.

Now try this

White cats are deaf. Would the Hardy–Weinberg equilibrium apply to white cats?   **(3 marks)**

Had a look ☐   **Nearly there** ☐   **Nailed it!** ☐

A level

Topic 5

Speciation

In order for a new species to form, part of an existing population must become reproductively isolated from another part.

Reproductive isolation

When two (or more) parts of a population cannot interbreed, **reproductive isolation** occurs. This can happen in a number of ways.

- Geographical isolation – populations are divided by a physical barrier, such as a river or mountain range.
- Habitat/ecological isolation – populations occupy different habitats in the same area so do not meet to breed.
- Seasonal/temporal isolation – species exist in the same area but are active for reproduction at different times.
- Mechanical isolation – the reproductive organs no longer fit together.
- Behavioural isolation – populations do not respond to each other's reproductive displays.
- Gametic isolation – male and female gametes from two populations are incompatible with each other.

Reproductive isolation can lead to allopatric and sympatric speciation.

Allopatric speciation

New species forming due to populations being physically or geographically isolated, is known as **allopatric speciation**.

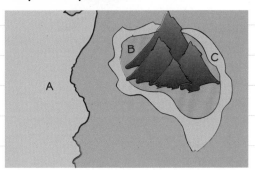

Population A could become geographically isolated from B if the island drifted off from the mainland, or sea level rose.

C could become isolated from B if a mountain range formed.

In both cases, interbreeding becomes impossible.

A change in allele frequencies due to natural selection (see page 59) in one of the isolated populations, could lead to allopatric speciation.

Isolation reduces gene flow between populations.

Sympatric speciation

Less commonly, two parts of a population can become isolated but be in the same place. This could lead to **sympatric speciation**.

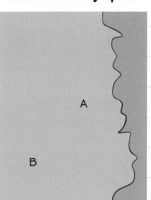

Here the two parts of the population, A and B, are separated but there is no physical barrier.

There may be habitat differences or some other mechanism that eventually leads to reproductive isolation.

An example of sympatric speciation can be seen in plants where a mistake in cell division leads to a change in chromosome number. The new and old forms cannot interbreed due to this difference and thus speciation has occurred instantaneously and in the same place. In animals this situation is much less common, but it has been seen happening in apple maggots (*Rhagoletis pomonella*) in the USA.

Worked example

Explain what happens during speciation. **(3 marks)**

Some sort of physical mechanism, such as a mountain range or a river, separates two groups of a population. If the two groups change in different ways, due to natural selection, they may become reproductively isolated and therefore separate species.

Now try this

A new species of mosquito has evolved in the London underground. Explain whether this is sympatric or allopatric speciation. **(2 marks)**

Death, decay and decomposition

Decay and decomposition are vital for life on Earth. Plants need nutrients such as nitrogen, potassium, phosphorus and carbon to make biomass. The extent of decay of a body, along with other features, can be used to determine the time of death.

Decay and decomposition

- Nutrients are locked into plant and animal tissues.
- Decay/decomposition processes release nutrients from dead plants and animals.

- Decomposition allows nutrients to be recycled.
- Microorganisms are crucial to decomposition.

The carbon cycle (see page 86) shows how nutrients are recycled, aided by microorganisms.

Time of death

There are various ways of telling what time someone died:

- **rigor mortis** – muscles stiffen and become fixed within 6–9 hours of death because contraction relies on ATP; this wears off again after about 36 hours

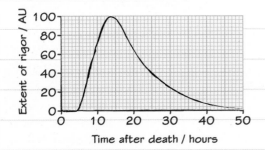

- **extent of decomposition** – bodies follow a standard pattern of decay; enzymes in the gut break down the wall of the gut and then the surrounding area; as cells die, they release enzymes, which break down tissues; the discoloration of the skin, and gas formation, combined with information about environmental conditions allow time of death to be estimated

- **forensic entomology** – finding the age of any insect larvae on the body allows the time the eggs were laid to be determined; an estimate of time of death can be made; there is a succession of species of insects on the body; the species present when the body is found allows the **stage of succession** to be determined and time of death estimated; the type of species found on a body can help identify the location of death

- **body temperature** – the body begins to cool after death, so temperature is used to find how long ago the person died; this is true for the first day or two; body size, position, clothing, air movement, humidity and surrounding temperature affect rate of cooling.

Worked example

The diagram shows the life cycle of a blowfly at 21°C.

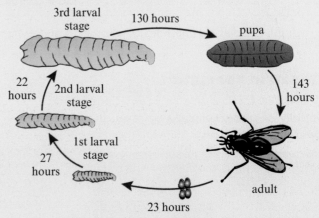

(a) Calculate the length of the cycle at 31°C assuming that all processes double in rate with a 10°C rise in temperature. **(2 marks)**

Time at 21 °C is the sum of the times for each stage which equals 345 hours. At 31 C it would take half this time, 172.5 hours. At 31 °C each process will take half the time so it will be $\frac{345}{2}$ hours = 172.5 hours

(b) Explain how a time of death estimate would be affected if investigations showed that a room had been at 31°C during the last week, although it was at 21°C when the body was found. This was due to a door being left open by the discoverer who found the body 2 hours ago. Adult flies were seen to be starting to emerge from pupae. **(2 marks)**

At 21 °C adults would be starting to emerge after about 14 days. At 31 °C it would be about 7 days. The original time of death estimate would be about 2 weeks ago but it would really be only 1 week.

Now try this

Analyse the information in the rigor mortis graph above to explain the relationship between time after death and extent of rigor. **(3 marks)**

DNA profiling

Genetic fingerprinting (**DNA profiling**) is used in forensic science. It is also used to identify and determine genetic relationships between organisms.

Everyone is unique

Everyone's DNA is unique apart from identical twins. This is because of the variety found in the sections of DNA that are not used to code for proteins. These non-coding sections are called **introns**. Short, repeated sequences of bases are called **short tandem repeats** (STRs). There can be several hundred copies of the STR at a single locus. People (and all other organisms) vary in regard to the number of these repeats they carry at each locus. Scientists look at the short tandem repeats at many loci to build up a unique pattern for that individual.

The **polymerase chain reaction** (PCR) allows tiny samples of DNA to be amplified so that they can be used in DNA profiling. The process relies on DNA primers, which are short sequences of DNA complementary to the DNA adjacent to the STR. A cycle of temperature changes results in huge numbers of the DNA fragments being produced.

	Reactants placed in vial in PCR machine
1	90–95 °C for 30 seconds – DNA strand separates.
2	50–60 °C for 20 seconds – primers bind to DNA strands.
3	72 °C for at least one minute – DNA polymerase builds up complementary strands of DNA
4	Steps 1–3 are repeated

1–3 repeated many times give millions of identical DNA copies to work with

---

🧪 Practical skills — Core practical 14 – DNA and RNA sequencing

DNA and RNA sequencing is a very important technique used to make a DNA profile.

The method of gel electrophoresis

xxxxxxxxxxx
Double-stranded DNA + restriction endonucleases

DNA is cut into fragments. →

xxxx xxx xx
Fragments of double-stranded DNA are loaded into the wells of an agarose gel tank.

The negatively charged DNA moves towards the positive electrode. The fragments separate into invisible bands. →

DNA is transferred to a nylon or nitrocellulose membrane by solution drawn up through the gel. DNA double strands split and stick to the membrane. →

Membrane placed in bag with DNA probe. Single-stranded DNA probe binds to fragments with a complementary sequence. →

If the DNA probe is radioactive, X-ray film is used to detect the fragments. If the DNA probe is fluorescent it is viewed using UV light.

In **core practical 14**, you are expected to use **gel electrophoresis** to separate DNA fragments of different length. In this process, DNA is cut into fragments with restriction endonucleases, which are separated. The result is called an **electrophoretogram** which is the DNA profile. Comparison of profiles between tissue samples from crime scenes and samples from suspects can help to identify perpetrators of crimes. Similarly, the technique can be used to look at how closely related individual plants and animals are, and help with determining evolutionary relationships.

Worked example

Following a burglary, a DNA profile was created from blood left on a broken window pane. The DNA profile was compared with those of four suspects, S1, S2, S3 and S4 (shown opposite). Analyse the data to deduce which of the suspects is likely to have left the blood sample on the broken windowpane.   **(4 marks)**

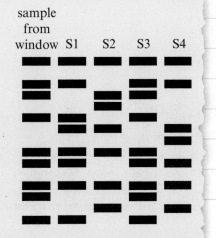

Suspect 3 matches all nine of the bands in the sample. DNA profiling assumes every individual's DNA is unique, apart from identical twins. DNA profiling analyses the non-coding short tandem repeats and non-coding DNA is very variable.

Now try this

State the names and functions of two enzymes used in DNA profiling.

(4 marks)

Exam skills

This exam-style question uses knowledge and skills you have already revised. Have a look at pages 82 and 83.

Worked example

(a) *It has been suggested that a reduction in the use of fossil fuels is necessary if further global warming is to be prevented. Explain why some scientists do not agree that a reduction in the use of fossil fuels will prevent further global warming.

(6 marks)

Carbon dioxide is produced by using fossil fuels, however, there is no direct evidence that increased carbon dioxide leads to global warming. In addition, carbon dioxide is released from other processes.

Removal of carbon sinks, such as rainforests, leads to increase in carbon dioxide.

Global warming is caused by gases, such as CFCs and methane, which do not come from fossil fuels. Methane, for example, comes from ruminant animals, paddy fields, melting ice and clearance of peatlands.

Natural cycles, such as in solar activity, may be involved in global warming.

Many of the deductions about global warming in the future rely on evidence from the past, which may not be an indicator of future events.

Some scientists may be biased, e.g. they are employed by a company with a vested interest in holding such views.

(b) First generation biofuels are made from sugars and vegetable oils found in food crops. Second generation biofuels are now being developed. These will use non-food parts of crops that contain the polymers cellulose and lignin. The graph below shows how the global production of first generation and second generation biofuels could change in the future. Analyse the data in the graph to explain the reasons for these changes.

(4 marks)

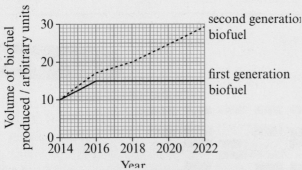

First generation biofuels level off after 2016. This is because they are made from edible components and further increases affect the food supply. Second generation biofuels continue to increase as they do not affect food supply because they are made using the non-edible components.

This question is unusual in that it requires a discussion of arguments **against** the commonly accepted theory of anthropogenic (human caused) global warming. It shows that the same knowledge can be required to be used in a variety of ways. A question asking about evidence for a reduction in the use of fossil fuels would rely on a lot of the same general ideas, just expressed differently. The question will require a consideration of more than one possible line of argument. The possible lines of argument are:

- currently there is a correlation but no evidence of causality
- CO_2 is produced from sources other than fossil fuels
- global warming is caused by other gases than CO_2
- there are natural processes which could be invoked to explain any temperature rise that is seen
- the past cannot predict the future
- the scientists who hold these views may be biased.

It is very unlikely that all these arguments would explain any one scientist's scepticism. This means that before planning an answer it would be important to notice that you are explaining the reasons why scientists rather than just one scientist might hold such views.

Make sure you structure your answer logically, showing how the points you made are related or follow on from each other where appropriate. You should also support your points with relevant biological facts and/or evidence.

Analysing data to **explain** means **two** things have to be done: in this case a description of the patterns shown by the data, followed by an explanation of the reasons for those patterns.

Bacteria and viruses

Bacteria and viruses are very different, but both cause disease in humans and other animals and plants.

Differences between bacteria and viruses

Bacteria	Viruses
cell surface membrane, cytoplasm, cell wall, ribosomes, plasmids and sometimes mesosomes, flagellum and pili see diagram on page 42	no cell wall, cell surface membrane, cytoplasm or organelles. nucleic acid enclosed in protein coat protein coat nucleic acid (DNA or RNA)
circular strand of DNA is the genetic material	DNA or RNA can be the genetic material
can live independently	must have a living organism as host
average diameter 0.5–5 µm	20–40 nm, wide range of sizes and shapes
often have mucus-based outer capsule	may have outer membrane of host cell surface membrane, but containing glycoproteins from the virus

TB, a bacterial disease

- *Mycobacterium tuberculosis* is carried in droplets when someone coughs or sneezes.
- The first infection may have no symptoms but tubercles form in the lungs due to the inflammatory response of the person's immune system.
- Some bacteria may survive inside the tubercles, due to their thick waxy coat. They lie dormant, but can become active again.
- Lung tissue is slowly destroyed by the bacteria, causing breathing problems.
- The patient develops a serious cough, loses weight and appetite and may suffer from fever.
- TB bacteria also target cells of the immune system.
- In some cases, the bacteria invade glands and the central nervous system (CNS). All of this can be fatal.

HIV/AIDS, a viral disease

- HIV/AIDS is in blood, vaginal secretions and semen and can be transmitted by needle sharing, unprotected sex, direct blood/body fluid transfer through cuts and from mother to foetus.
- The initial symptoms are fevers, headaches, tiredness and swollen glands but sometimes no symptoms.
- Three to twelve weeks after infection, HIV antibodies appear in the blood, and the patient is now HIV positive.
- All symptoms can then disappear for years but eventually patients suffer from weight loss, fatigue, diarrhea, night sweats and infections such as thrush. This is now thought of as full-blown AIDS.
- Finally dementia, cancers (e.g. Kaposi's sarcoma) and opportunistic infections such as TB arise and may lead to death.

Worked example

Explain why a patient infected with TB is more likely to develop symptoms of the disease if HIV also infects them. **(3 marks)**

HIV destroys T helper cells. This means bacterial cells are not destroyed by the immune system and so they proliferate and lead to TB.

Now try this

The antibiotic penicillin acts on a bacterium by weakening its cell wall so that it bursts. The antibiotic tetracycline prevents the production of new proteins by the bacterium. State **one** reason why each antibiotic is not effective against viruses.

(2 marks)

Pathogen entry and non-specific immunity

Your body is under constant threat of invasion from **pathogens** (organisms that cause diseases) but you have numerous lines of defence.

Preventing entry, the first line of defence

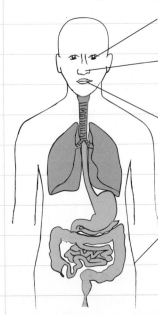

eyes – tears contain the enzyme lysozyme which helps to digest microbes.

respiratory tract – contains mucus which traps bacteria. The mucus is then swallowed and passed into the digestive system.

gastrointestinal tract – acid in the stomach helps to protect against any microbes which are eaten. In addition the gut has its own bacteria. These compete with pathogens for food and space which helps to protect us. The harmless bacteria also excrete lactic acid which deters pathogens.

skin and skin flora – the skin is a tough barrier and usually only allows pathogens to enter if it is cut. As an additional line of defence the skin has its own microbes. These live naturally on the skin and out-compete pathogens. Sebum is an oily fluid which is made by the skin and can also kill microbes.

How do pathogens enter the body?

They enter through areas not covered by skin: nose, mouth, gas exchange surfaces, the eyes, gastrointestinal tract and genital tract. The entry of microorganisms through wounds is also a major cause of infections.

Non-specific responses, the second line of defence

- **Inflammation** – damaged white cells release histamines that cause arterioles to dilate and capillaries to become more permeable; blood flow to the area increases and plasma, white blood cells and antibodies leak out into tissues, where they can attack the pathogen.

- **Lysozyme action** – an enzyme found in tears, sweat and the nose, destroys bacteria by breaking down the bacterial cell walls.

- **Interferon** – a chemical released from infected cells, which prevents viral replication.

- **Phagocytosis** – white blood cells engulf, digest and destroy bacteria and other foreign material; the foreign material is enclosed in a vesicle, into which the cell secretes digestive enzymes from lysosomes, which destroy the bacteria; these phagocytes include neutrophils and monocytes (which become macrophages), these cells are also involved in the specific immune response (see page 94).

Worked example

The skin has an important role in protecting the body from infection by pathogenic bacteria. Human skin has a community of microorganisms, called the skin flora, living on it. Most of these microorganisms are harmless bacteria that feed on dead skin cells and secretions.

Explain **two** ways in which the skin flora can help to protect a person from infection by pathogenic bacteria.   **(4 marks)**

The skin flora will compete with the pathogens for nutrients, therefore effectively 'starving' them. In addition, the skin flora is known to secrete chemicals which kill or reduce the reproductive rate of the pathogens, therefore reducing the population size.

 Explain questions require you to say **what** happens and then **why** this is useful. Words like 'therefore' and 'because' are very good to use here.

Now try this

Compare and contrast the production and action of interferon with that of lysozyme.

(3 marks)

Specific immune response – humoral

Both **humoral** and **cell-mediated** immune responses involve processes of **antigen presentation**.

The humoral response

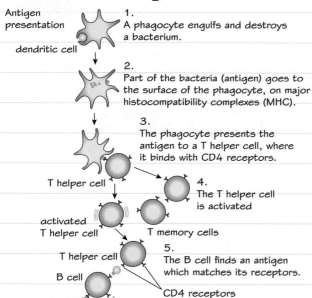

Antigen presentation

dendritic cell

1.
A phagocyte engulfs and destroys a bacterium.

2.
Part of the bacteria (antigen) goes to the surface of the phagocyte, on major histocompatibility complexes (MHC).

3.
The phagocyte presents the antigen to a T helper cell, where it binds with CD4 receptors.

T helper cell

4.
The T helper cell is activated

activated T helper cell

T memory cells

T helper cell

5.
The B cell finds an antigen which matches its receptors.

B cell

CD4 receptors

bacteria   cytokines

6.
It is activated by a T helper cell which binds to it and produces cytokines.

10.
If the same intruder invades again, memory cells help the immune system to activate much faster.

7.
Then the B cell divides to produce plasma and memory cells.

plasma cell

memory cell

plasma cell

B memory cell

bacteria

antibodies

8.
Plasma cells produce antibodies that attach to the current type of invader.

9.
Macrophages identify intruders marked with antibodies, and engulf and destroy them. Antibodies also clump the bacteria together, which helps to stop them spreading and they can neutralise toxins.

Lymphocytes

B lymphocytes (B cells) are produced and mature in bone marrow. They target microorganisms and soluble toxins in the humoral response. Some B cells become **plasma cells** and secrete antibodies which cause removal of antigens.

T lymphocytes (T cells) are produced in the bone marrow, and mature in the thymus gland. They include:

* T killer cells destroy body cells that have foreign proteins on their surface, those infected by a virus or abnormal cells in cancers.
* T helper cells stimulate all specific responses.

Antigens

An **antigen** is anything that causes an immune response; this is usually foreign proteins or glycoproteins. Each protein may have a number of antigenic regions. A whole organism such as a bacterium will carry many antigens.

Antibodies

Antibodies work in a variety of ways to rid the body of antigen-carrying structures, such as microorganisms. Some ways include:

* **opsonisation** – particles are coated with antibodies, marking them for phagocytosis
* **precipitation** – soluble toxins are made insoluble and inactive
* **agglutination** – microorganisms are clumped together for easier phagocytosis
* **lysis** – breaking open of bacterial cells.

Infection with HIV, a virus, can result in a condition called AIDS in which the immune system does not function. HIV specifically infects T helper cells. Explain how this will result in the loss of the immune response.   **(4 marks)**

Viral infection will cause the cell-mediated response to the infected cells.

T helper and T killer cells will become activated and the T killer cells will destroy infected T helper cells.

Loss of T helper cells will prevent further activation of T killer cells, so the cell-mediated response will not function.

T helper cells are also needed to activate B cells in the humoral response, hence this immune response will also be lost.

 The role of T helper cells in both types of immune response is important here.

Now try this

Describe the differences in the roles of B cells and T cells as effector cells in the removal of antigen from the body.   **(3 marks)**

Specific immune response – cell-mediated

When pathogens enter body cells, the humoral response is unable to get to them. In this case, the **cell-mediated** response is used to get them out of the cell.

Cells of the cell-mediated immune response

- Body cells act as **antigen-presenting cells** (APCs) when infected.
- **T helper cells** with specific surface receptors bind to the APCs and are activated.
- Specific **T killer cells** also bind to surface antigen of APCs.
- **Activated T helper cells** produce cytokines to activate T killer cells.
- Activated T killer cells become cytotoxic, killing cells showing the antigen.
- **T memory cells** remain, ready to respond to further infection.

Similarities of humoral and cell-mediated responses

- Both are specific to a particular antigen.
- Both require formation of APCs and the presence of **major histocompatibility complex** (MHC).
- Both involve **clonal selection**, which is the activation of only those cells that recognise the antigen.
- Both involve **clonal expansion**, which is proliferation of the cells needed so that there are very large numbers.
- Both involve differentiation of cells, following activation by T helper cells, to produce **effector cells**.

Cell-mediated response

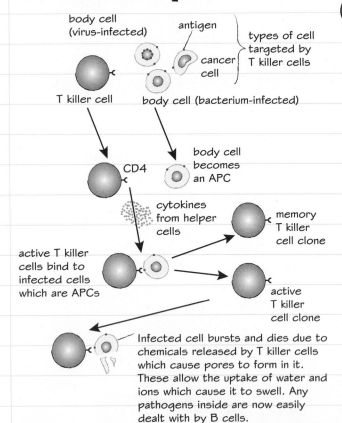

active T killer cells bind to infected cells which are APCs

Infected cell bursts and dies due to chemicals released by T killer cells which cause pores to form in it. These allow the uptake of water and ions which cause it to swell. Any pathogens inside are now easily dealt with by B cells.

The cell-mediated response that kills body cells releasing pathogens, and also kills cancer cells. Unfortunately, this response also kills cells in transplanted organs which is why the immune system has to be suppressed after a transplant.

Worked example

Compare and contrast the humoral and cell-mediated immune responses.   **(5 marks)**

Both are specific to a particular antigen. Both require formation of APCs and the presence of MHC. Both involve clonal selection, which is the activation of only those cells that recognise the antigen. Both involve clonal expansion, which is proliferation of the cells needed so that there are very large numbers. Both involve differentiation of cells, following activation by T helper cells, to produce effector cells.

The humoral response is to particles and solutes, cell-mediated is to infected body cells.

The humoral response produces antibodies, cell-mediated produces T killer cells.

Compare and **contrast** will always occur together and requires similarities and differences. At least one example of each will be needed for full marks. Always give both halves of a comparison, even if you're just saying ' ... and the other doesn't'.

Now try this

Describe the role of T helper cells in the immune response.   **(3 marks)**

Post-transcriptional changes to mRNA

Until recently, it was thought that one gene coded for one protein. It is now known that the mRNA made in transcription is modified to create a number of mRNAs, which code for different proteins.

mRNA splicing

One gene can code for more than one protein. This is achieved by **post-transcriptional changes** in the mRNA:

- the mRNA made in the nucleus is pre-mRNA
- non-coding regions, called **introns**, are removed by spliceosomes
- the coding regions (**exons**) are then spliced together to form the mRNA
- the exons can be **spliced** in different combinations to give different mRNAs and thus different proteins.

Intron removal

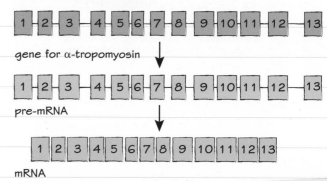

gene for α-tropomyosin

pre-mRNA

mRNA

The gene shown has 13 exons (red blocks) and 12 introns (black lines). When pre-mRNA is made, all regions are copied (transcribed) but on production of the **mature** mRNA the introns are removed.

RNA splicing

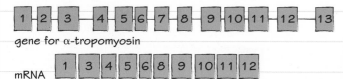

gene for α-tropomyosin

mRNA

OR

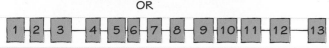

gene for α-tropomyosin

mRNA

The gene for α-tropomyosin (shown in the 'intron removal' diagram above) is the starting gene here too. The introns are removed, and this time so are some of the exons. The exons removed are different in each case, leading to two different mature mRNAs that would translate to make two different proteins.

Worked example

The frequency of sound detected in the inner ear of chickens depends on the type of BK channel protein present in the hair cell membrane. There are thought to be 48 different BK channel proteins in these membranes. The *cSlo* gene codes for all of these BK channel proteins.

Explain how one *cSlo* gene can give rise to different BK channel proteins in these hair cells.   **(5 marks)**

The DNA is transcribed to make pre-mRNA.

This pre-mRNA is then modified by the removal of introns.

This is carried out by spliceosomes.

The exons are rejoined in different combinations.

The mRNAs produced are different and are translated to produce different sequences of amino acids, which results in different bonds stabilising the protein's tertiary structure.

Now try this

The diagram shows a portion of mRNA with exons (white) and introns (grey). Draw out the amino acid sequence you would find in the protein made from this mature mRNA.   **(3 marks)**.

U G C C A UCGGAUCUGUGUU GUAAAUUG C G G C G G A U UUGCCACGAA A C U A G C

97

Types of immunity

If someone has been exposed to a pathogen in the past their immune response will be faster the next time they encounter it. This ability can be acquired naturally or by vaccination. In the short term, a person at risk of infection can be protected passively.

Secondary immune response

Memory cells produced after first exposure to an antigen (the **primary response**) allow a faster and greater response to future exposure (**secondary response**) so fewer symptoms occur.

Benefits of a secondary response

- ✓ Much shorter lag period as more of the specific lymphocytes in circulation.
- ✓ More rapid production of effector cells.
- ✓ Much greater production of antibody or T killer cells.

Primary and secondary immune responses

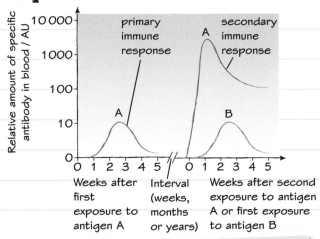

Note that the secondary response is specific to a particular antigen (A in this graph). Antigen B stimulates a primary response.

How immunity is developed

	Active body cells involved	Passive protection developed in other individual
Natural normal body response	normal response to foreign antigen (look back at page 95)	antibodies cross the placenta to protect the baby while its immune system develops; breast milk, especially colostrum, also contains antibodies
Artificial medical procedure	vaccines use antigens from pathogens in a harmless form to cause a primary response; reinfection will then cause a secondary response	an antiserum contains antibodies, usually from the blood of animals, following exposure to an antigen; injecting these antibodies will neutralise the antigen, but will not provide long-term protection

Worked example

Streptokinase is an enzyme that can be injected into patients to digest blood clots that have formed in their blood vessels. The graph shows measurements of antibodies produced against the enzyme following injection. Explain why frequent use of streptokinase to control blood clots is not possible. **(3 marks)**

Streptokinase is a protein that acts as an antigen. First injection results in a primary immune system response, the second in a secondary response. This acts as a vaccine and hence antibodies from circulation will remove streptokinase.

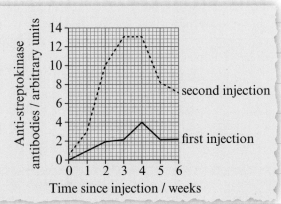

Now try this

Explain why antibodies provided by a mother for her baby in her breast milk give only short-term protection whereas a vaccine can last for many years. **(4 marks)**

Consider the type of immunity each of them provides.

Antibiotics

 Practical skills

It is not possible to vaccinate against all infections, and if the immune system does not respond quickly enough, an infection may be lethal. **Antibiotics** can kill or slow down the pathogen.

Types of antibiotic

There are two kinds of antibiotics to fight bacterial infections:

1 **bactericidal** – an antibiotic that kills bacteria

2 **bacteriostatic** – an antibiotic that limits or slows the growth of bacteria.

When bacteria are no longer affected by an antibiotic they are said to be resistant to it (see page 99).

Investigating bacteria and antibiotics – Core practical 15

Core practical 15 requires you investigate the effect of different antibiotics on bacteria.

A simple procedure can be carried out to investigate bacteria and antibiotics for the core practical. It is important to follow safe, aseptic techniques (for a reminder of aseptic techniques, see page 68) when doing this kind of work, and it is essential to carry out a risk assessment.

> A sterile nutrient agar plate is seeded with suitable bacteria, e.g. using a *sterile* spreader.

↓

> Apply antibiotic to a *sterile* filter paper disc, then lay on the bacterial lawn using *sterile* forceps.
> OR
> Place antibiotic solution into a well in agar, using a *sterile* pipette.

↓

> Seal Petri dish but do not seal all around the dish.

↓

> Incubate below 30°C for about 24 hours.

↓

> Look for clear areas around the antibiotic discs or wells. Bigger areas indicate a better antibiotic against this bacterium species.

Evaluate questions ask you to identify the relevant information from data and then reach a conclusion. Here, the conclusion is what you can work out about possible treatment for AIDS patients.

Worked example

The results of a study on the effects of two different antibiotics on populations of bacteria are shown in the graphs. Analyse the data to evaluate the use of the two antibiotics to treat bacterial infection in individuals with AIDS who have an ineffective immune system.

(4 marks)

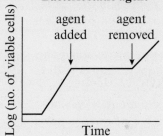

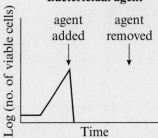

Bactericidal antibiotic kills the bacteria, shown by the population falling to zero.

Population does not recover when antibiotic removed so treatment would be effective.

Bacteriostatic antibiotic halts growth of the population as the curve becomes horizontal.

Live cells remain in the population and divide again when the antibiotic is removed.

For AIDS patients, a bactericidal antibiotic would be effective with a course of treatment, bacteriostatic antibiotics would need to be taken continuously.

Now try this

Describe in outline how you could test the effectiveness of an antibiotic on a specific bacterium in the laboratory. Include aspects of the method that ensure safe working.

(5 marks)

Evolutionary race

Pathogens (disease-causing organisms) are said to be in an evolutionary race with humans and their medicines (for reminder of niches and how organisms can adapt, see page 58).

Hospital-acquired infection

Mutations help some bacteria to become resistant to antibiotics, which results in problems with antibiotic-resistant bacterial infections in hospitals. Hospitals try to combat this in a number of ways through their **code of practice**.

Hospital code of practice	How it helps to stop antibiotic resistance
only use antibiotics when needed and ensure course of treatment is completed	reduces selection pressure on organisms and destroys all bacteria causing infection
isolating patients with resistant diseases	prevents transmission of resistant bacteria between patients
good hygiene encouraged, including hand washing and bans on wearing of jewellery, ties and long sleeved shirts	prevents the spread of infection and cuts down on the number of places that may harbour pathogens
screening of patients coming into hospital	a person may be infected without showing symptoms; this can be detected and they can be isolated and treated

Evading the immune system

The human immune system is one of the main selection pressures on pathogens. However, there has been an evolution of bacterial strategies for evading the immune system. Slight changes in the pathogen's antigens mean that memory cells from a previous infection will be useless in combating a second infection.

Antibiotic resistance

Pathogens are said to be in an evolutionary race with humans and their medicines. From the mid-1940s penicillin and other antibiotics combated bacterial infections, and antibiotics were prescribed widely. Certain bacteria (such as *Staphylococcus aureus*) quickly mutated and became resistant to penicillin. This cycle continues with scientists developing new drugs and bacteria evolving.

Worked example

Hospital-acquired infections caused by bacteria can be a major problem for patients. In a study in a London hospital, it was found that pillows contaminated with bacteria could spread infections between patients. Explain how this hospital could improve the prevention and control of the spread of infections. **(3 marks)**

The hospital should change its code of practice for dealing with hospital-acquired infections. This should include better clothing rules for hospital workers, higher washing temperature and the treatment of pillowcases with antibacterials. Bed linen should be carried to the laundry in sealed plastic bags. The code should include a highly publicised and implemented hand-washing regime.

Now try this

A new strain of HIV has been found in people who are resistant to HIV. This strain is very similar to previous strains, but has genetic differences that prevent it from being destroyed by their immune systems. However, the virus is unable to replicate properly which means that only small numbers are present in the bloodstream and AIDS does not develop.

Explain how this is an example of an 'evolutionary race' between HIV and humans. **(4 marks)**

Exam skills

This exam-style question uses knowledge and skills you have already revised. Have a look at page 73 to review the information.

Worked example

1 A method that can help with the problem of distinguishing between species is DNA sequencing/profiling. The diagram shows the results of running DNA sequencing on a digested gene from six species of clown fish (lanes 1–6), and the undigested protein (lane 7). Analyse the information to comment on how far the sequencing data confirm the current classification. **(5 marks)**

> The command word **comment** means the using a number of variables from data/information to form a judgement.

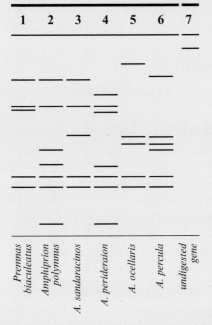

> How much to write – there is a lot to say here and if you analysed it all in full detail it could stretch to a whole page or more. Instead, look at the number of marks to be awarded. In this case it is 5, so you must make up to 5 points. Here, the answer refers to lane 1 and the other lanes (1), then is followed by a supporting point for this (1). A point about the genus *Amphiprion* (1). A general point about the difference shown in all species (1). A point about the two possibly most closely related species and expansion of this (2).

Lane 1 is currently classified in a different genus from all the others but its sequence shows no obvious bigger difference, giving no support to it being in a different genus. It has five bands, as do lanes 3 and 6 of a different genus. By the same token, the *Amphiprion* species have no obvious similarity to each other which gives no support to them being in the same genus. However, the patterns of bands are different for all six species studied, so this supports the existing view that they are separate species. The data may suggest that *A. polymnus* and *A. perideraion* are the most closely related as they are the only ones with the lowest band, and of the other six bands, they have four in common.

2 The genetic relationship between two species of grey tree frog has been studied using DNA profiling.
   Describe how these DNA profiles would be compared.

> This question shows the importance of reading what is required very carefully. It does not ask how the profiles would be made but how they would be **compared**.

The profiles would be compared with respect to the total number of bands, the position of the bands and the size of the bands.

3 Describe how small DNA samples could be amplified.

> Again, it is important to read the question. This is about amplification not actual profiling. You should make sure you know the details of such processes.

Multiple copies of the DNA can be made using a technique called the polymerase chain reaction. In this process, the DNA is mixed with primers, DNA polymerase and nucleotides. This mixture is then held at 90 to 95 °C, then at 50 to 65 °C and finally at 70 to 80 °C. The cycle is repeated many times.

Skeleton and joints

The bones of the skeleton are attached to each other by **ligaments** and to muscles by **tendons**. These four components interact to create movement.

Bones and joints

- Bones can move in relation to one another at joints.
- Different types of joint allow different degrees of movement.
- Ligaments are made of elastic connective tissue and hold bones together, and restrict the amount of movement possible at a joint.
- Tendons are cords of non-elastic fibrous tissue that anchor muscles to bones.

Muscles

- Skeletal muscles are those attached to bones and are normally arranged in **antagonistic pairs**; this means pairs of muscles which pull in opposite directions.
- **Flexors** contract to flex, or bend, a joint, e.g. biceps in the arm.
- **Extensors** contract to extend, or straighten, a joint, e.g. triceps in the arm.

Synovial joint – the knee

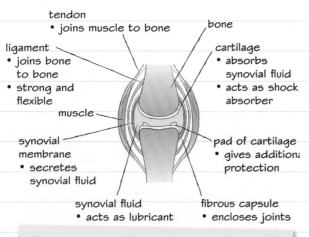

tendon
- joins muscle to bone

bone

ligament
- joins bone to bone
- strong and flexible

muscle

synovial membrane
- secretes synovial fluid

cartilage
- absorbs synovial fluid
- acts as shock absorber

pad of cartilage
- gives additional protection

synovial fluid
- acts as lubricant

fibrous capsule
- encloses joints

There are a number of types of joint but the synovial is the most typical; this is the knee.

Types of joint

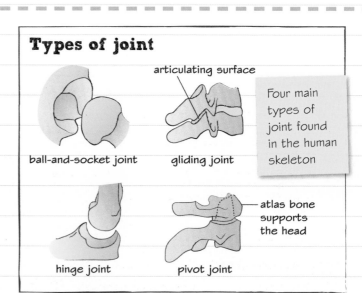

articulating surface

ball-and-socket joint

gliding joint

hinge joint

pivot joint

Four main types of joint found in the human skeleton

atlas bone supports the head

Worked example

The diagram shows some of the muscles in the human leg. Explain how these muscles would interact in order for the person to gently kick a ball. **(3 marks)**

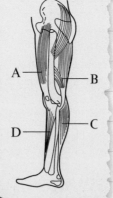

Muscle B would contract to pull the lower part of the leg back; this is flexion. Muscle A would then contract to pull the lower part forward to kick the ball; this is extension.

Always read all the words in a question. Here, the word **gently** is significant as it tells you that there is no running or vigorous kick, so the two muscles A and B will be enough to swing the lower leg back and then forward again.

Also, remember that muscles can't stretch themselves. It is the pull created by the contraction of the antagonistic muscle that stretches the other muscle in the antagonistic pair when it is in a relaxed state.

Now try this

Identify precisely structures A–D in the elbow joint. **(4 marks)**

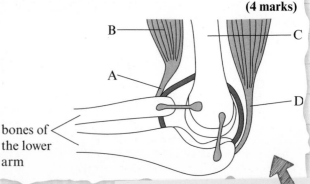

bones of the lower arm

Note the word **precisely** in this question. A good idea in an exam is to look carefully at questions and highlight key words like this. The examiner will not do that for you!

Muscles

The **sliding filament theory** of muscle contraction requires a clear understanding of muscle structure. The filaments that slide are the proteins **actin** (thin) and **myosin** (thick). Another protein, tropomyosin, is held in place by troponin, and blocks contraction until calcium ion concentration increases. The increase in calcium ion concentration stimulates the muscle to contract.

The sarcomere

- The functional unit of a muscle fibre (also called a muscle cell) is called a **sarcomere**.
- There are numerous sarcomeres in a muscle cell.

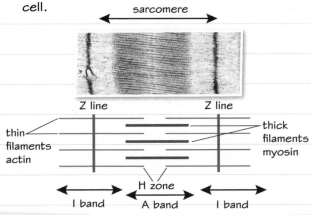

A sarcomere

Look at the vertical blue lines to see the shortening that has happened between A and B.

The sliding filament theory

When the muscle contracts the thin actin filaments move between the thick myosin filaments, shortening the length of the sarcomere and therefore shortening the length of the muscle.

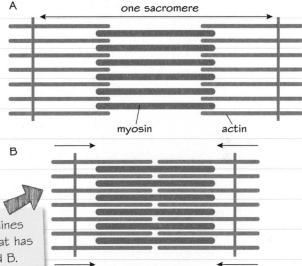

How it works

Muscle contraction is based on many coordinated movements happening at the same time, and then repeating again and again. The whole process is started by the arrival of a nerve impulse that causes the release of calcium ions.

Sequence of events in muscle contraction. This is repeated in many sarcomeres and time and time again in each one.

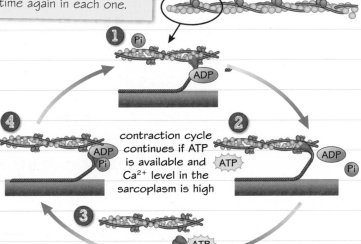

contraction cycle continues if ATP is available and Ca^{2+} level in the sarcoplasm is high

Use the diagram opposite to complete the flow chart showing the events of muscle contraction. **(6 marks)**

Calcium attaches to the troponin molecule, causing it to move → tropomyosin moves, exposing myosin-binding sites → myosin heads bind to them to form cross-bridges → ADP and Pi on the myosin head are released → myosin changes shape and its head nods forward → the filaments move due to this → the actin moves over the myosin → ATP binds to the myosin head causing the myosin head to detach → ATPase on the myosin head hydrolyses the ATP to ADP and Pi → this causes a change in the shape of the myosin head and returns to its upright position → the bending of many myosin heads combines to move the actin filaments relative the myosin filament.

Now try this

Explain the role of calcium ions and ATP in muscle contraction. **(5 marks)**

The main stages of aerobic respiration

Cellular respiration provides **ATP** for processes within a cell. **Aerobic respiration** has three main stages (**glycolysis, Krebs cycle, electron transport chain**) that result in energy being transferred from **glucose** to ATP when oxygen is available.

Overall result of aerobic respiration

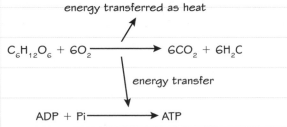

$$C_6H_{12}O_6 + 6O_2 \longrightarrow 6CO_2 + 6H_2O$$

energy transferred as heat

energy transfer

$$ADP + Pi \longrightarrow ATP$$

About 30 molecules of ATP are produced from one molecule of glucose.

Remember that energy can be changed or transferred – it cannot be created or produced.

Carbon dioxide is released as waste and hydrogen is reunited with atmospheric oxygen.

The ATP/ADP cycle

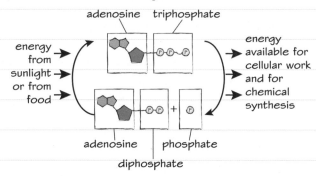

Energy is eventually transferred as heat to the energy store of the surroundings.

ATP – the universal energy currency

ATP is constantly produced and used within cells as a source of energy for metabolic processes. ATP is produced by respiration in all organisms and also by photosynthesis in plants. Processes within cells that require energy, break down ATP to ADP and Pi and energy is transferred.

How the stages of aerobic respiration work together

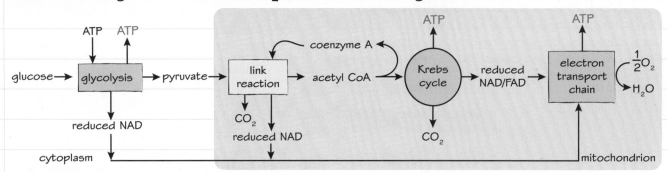

Some ATP is made directly in glycolysis and the Krebs cycle (substrate level phosphorylation). The rest is made when reduced NAD and FAD carry H⁺ and electrons to the electron transport chain (ETC). NAD and FAD are coenzymes that are reduced when hydrogen atoms are removed from molecules during reactions that break down glucose in glycolysis, the link reaction and the Krebs cycle. Each step in each of these processes is catalysed by a specific enzyme found in the cytoplasm (glycolysis and link) or mitochondria (Krebs and ETC). These are all **intracellular enzymes**.

Worked example

A fairly active person uses about 75 kg of ATP during a day but there is only around 50 g in the body at any time. Explain how enough ATP can be supplied to the cells. **(2 marks)**

The ATP present is produced at a rate that is high enough to supply the cells. Using these figures, the cycle between ATP and ADP happens around 7500 ÷ 50 times during the day. That is 1500 times.

Now try this

Describe the roles of coenzymes in aerobic respiration. **(5 marks)**

The command word here is **describe**. Just list what the coenzymes do; there's no need to explain how or why.

Glycolysis

Glycolysis is the first stage of respiration and occurs in the cytoplasm of a cell, converting glucose to pyruvate.

What happens in glycolysis

- Glycolysis provides pyruvate for aerobic respiration in mitochondria and also for anaerobic respiration in the cytoplasm (see page 105).
- There are many small steps that result in glucose being split into two molecules of pyruvate.
- ATP is needed to make glucose more reactive at the start of the process but more ATP is produced than used.
- NAD is reduced and may be used to generate ATP in the mitochondrion if there is oxygen.

Glycolysis in outline

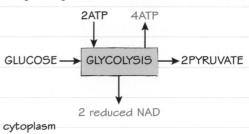

cytoplasm

Steps in glycolysis

 Phosphorylation

Two molecules of ATP are used to add phosphate to glucose.

2ATP   2ADP

GLUCOSE ⟶ HEXOSE BISPHOSPHATE

This results in hexose bisphosphate which is more reactive than glucose in preparation for the next steps in breaking down the glucose.

 Lysis hexose molecule

Hexose bisphosphate is broken into two molecules of glycerate-3-phosphate (GP).

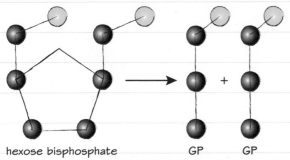

hexose bisphosphate   GP   GP

> Black circles are carbon, yellow are phosphate; H and O not shown

③ Formation of pyruvate

NAD is reduced to form NADH (reduced coenzyme) by removing hydrogen atoms from GP. This is dehydrogenation, and GP is oxidised as a result. Four molecules of ATP are formed using the energy released as bonds are formed.

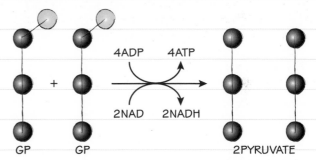

GP   GP   2PYRUVATE

These four ATP molecules give the net two ATP produced by glycolysis (two ATP were used in step 1).

The ATP is produced by substrate level phosphorylation which is different from the ATP production in the mitochondrion (oxidative phosphorylation).

The NADH will be used to produce more ATP in the mitochondrion.

Worked example

Explain that the energy for ATP production during glycolysis is provided by oxidation, despite the fact that oxygen is not required. **(4 marks)**

Removal of hydrogen atoms from GP is a form of oxidation (the removal of hydrogen or an electron). Energy is transferred to ATP molecules in substrate level phosphorylation.

Now try this

Explain the significance of the following in glycolysis.

(a) phosphorylation **(3 marks)**

(b) NAD **(3 marks)**

Link reaction and Krebs cycle

Pyruvate produced in glycolysis enters the matrix of the mitochondrion where the **link reaction** and the **Krebs cycle** occur. The glucose is completely oxidised.

Link reaction

- Pyruvate is oxidised as it loses hydrogen atoms to NAD. This will be used to form ATP.
- Decarboxylation (removal of carbon dioxide) occurs to form carbon dioxide waste.
- The remaining acetyl group (two carbon atoms) is combined with coenzyme A to form acetyl CoA.

Link reaction summary

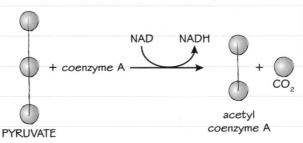

Two molecules of acetyl CoA are formed from each glucose molecule. Acetyl CoA combines with a 4-carbon (4C) compound in the Krebs cycle.

Krebs cycle

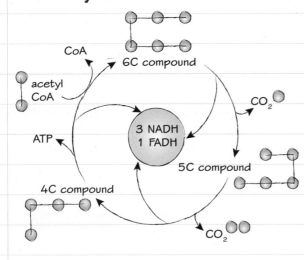

Significance of Krebs cycle

- Krebs produces reduced coenzyme (NADH and FADH) by oxidation of the molecules produced using acetyl CoA.
- Energy from reduced coenzymes will be used to produce ATP.
- Two carbons are added to the cycle by acetyl CoA. Later, two carbon atoms are removed to form carbon dioxide waste. This is decarboxylation.
- Coenzyme A is recycled to be used again in the link reaction.
- One molecule of ATP is produced per acetyl CoA by substrate level phosphorylation.
- The cycle reforms the 4C compound that combines with acetyl CoA.

Worked example

The diagram below shows some of the stages in the Krebs cycle. Production of reduced coenzyme has not been shown.

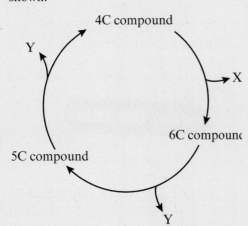

(a) Describe the formation of the 6C compound shown. **(2 marks)**

Two carbon atoms are added to the 4C acceptor molecule from acetyl coenzyme A. Coenzyme A is released during the reaction.

(b) Give the name of molecule X. **(1 mark)**

ATP

(c) Give the name of molecule Y. Justify your answer. **(2 marks)**

Carbon dioxide. The number of carbons in the cycle falls by one following each production of Y.

Now try this

Discuss the effect of a lack of acetyl coenzyme A on the Krebs cycle. **(3 marks)**

The **discuss** command word expects you to give reasoning in your answer.

The command word **justify** is used when an explanation is required for a statement in the question or one of your answers.

Note that NADH and FADH have been excluded by the stem of the question.

Oxidative phosphorylation

Most ATP produced by aerobic respiration is by **oxidative phosphorylation** in the inner membrane of the **mitochondrion**.

Oxidative phosphorylation

Phosphorylation is the addition of phosphate, in this case adding to ADP, and oxidative means that oxidation is the source of energy. Oxidative phosphorylation is the term for ATP production in the electron transport chain.

Electron transport chain (ETC)

1 NADH and FADH are oxidised by releasing electrons to proteins in the ETC and releasing H^+.

2 Electrons release energy as they are passed along a chain of proteins in the ETC.

3 Energy is used to pump H^+ from the matrix into the intermembrane space.

4 H^+ flows through proteins in the membrane, including the enzyme ATP synthase.

5 ATP is produced as the ions flow towards the matrix.

6 Electrons leaving the ETC are passed to an oxygen atom and combine with $2H^+$ to form water. Oxygen is known as the terminal electron acceptor.

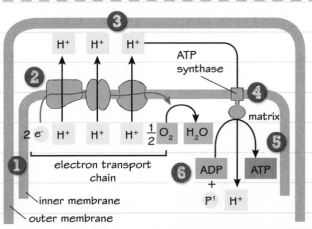

The events in the inner membrane of the mitochondrion

ATP is synthesised by chemiosmosis

The inner mitochondrial membrane separates the mitochondrial matrix from the intermembrane space. Also, important enzymes and the proteins of the ETC are embedded in the membrane. The ETC creates a diffusion gradient for H^+ and also a charge difference between the matrix and the intermembrane space by pumping H^+ across the membrane.

The inner membrane is not permeable to hydrogen ions so they can only flow into the matrix through a protein channel in the membrane called ATP synthase. The flow of ions provides the energy to convert ADP and phosphate into ATP.

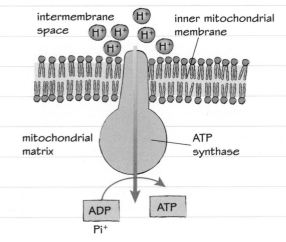

Worked example

Hydrogen ions from earlier stages of respiration are passed to the ETC in the mitochondrion.

(a) Name two stages of respiration that provide hydrogen atoms. **(1 mark)**

Glycolysis and Krebs cycle.

(b) Describe how hydrogen atoms are carried to the electron transport chain. **(1 mark)**

In the form of reduced coenzymes (NAD and FAD).

(c) Describe what is meant by oxidative phosphorylation. **(3 marks)**

The formation of ATP, using energy from oxidation, in the electron transport chain.

Now try this

Isolated mitochondria can be treated in order to make their membranes permeable to hydrogen ions. When provided with pyruvate these mitochondria are able to oxidise reduced NAD but they do not produce ATP.

Explain the observations regarding the functioning of these mitochondria. **(4 marks)**

Anaerobic respiration

ATP can still be produced in cells without oxygen by **anaerobic respiration**, but the efficiency of energy release from glucose is much lower than that of aerobic respiration.

Respiring without oxygen

- Without oxygen to use as the terminal electron acceptor, the electron transport chain cannot function.
- This will mean that no NAD is available to accept hydrogens in the Krebs cycle so the production of NADH is not possible.
- Glycolysis also requires NAD. Without oxygen, pyruvate is used to oxidise NADH and this makes NAD.
- The pyruvate is converted into **lactate**.

Anaerobic respiration is inefficient

Only **glycolysis** is involved in anaerobic respiration. Glucose is only partially broken down in anaerobic respiration to energy-rich molecules, meaning that only a fraction of the possible energy of glucose is used to make ATP.

Just two ATP molecules result from each glucose compared with around 30–32 in aerobic respiration.

Anaerobic respiration in animals

Lactate is a necessary end product of the oxidation of NADH to regenerate NAD so that glycolysis can continue. However, lactate becomes lactic acid and this lowers blood pH. This causes a reduction in enzyme activity and so the lactate must be removed. Most of it is converted back into pyruvate when oxygen is available again. The extra oxygen required to do this is called the **oxygen debt**.

Anaerobic respiration during exercise

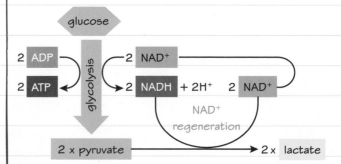

After exercise, a person breathes more deeply than they would normally do at rest to pay off the oxygen debt due to anaerobic respiration. Some of the lactate is made into glycogen and stored in the liver.

Worked example

Complete the table below to compare aerobic and anaerobic respiration in a mammalian muscle cell.

(6 marks)

Feature	Aerobic	Anaerobic
site in a cell	cytoplasm and mitochondrion	cytoplasm
final hydrogen acceptor	oxygen	pyruvate
products	CO_2 and H_2O	lactate
ATP required	2	2
ATP produced	30	2
method of ATP production	oxidative and substrate level phosphorylation	substrate level phosphorylation

Now try this

Fast twitch muscle fibres operate anaerobically most of the time. It is fast twitch fibres that provide muscles with the power needed for sprinting over short distances.

(a) *Explain how a sprinter is able to provide enough energy to run quickly without using oxygen. **(6 marks)**

(b) Explain why sprinting is limited to short distances. **(2 marks)**

Make sure you structure your answer logically, showing how the points you make are related or follow on from each other where appropriate.

The rate of respiration

Practical skills **Core practical 16** requires you to investigate the rate of respiration. The rate of aerobic and anaerobic respiration can be determined using a **respirometer** to measure the rate of oxygen absorbed or carbon dioxide produced.

The equation for respiration

The overall equation for aerobic respiration of glucose is: $C_6H_{12}O_6 + 6O_2 \rightarrow 6CO_2 + 6H_2O$

In theory, the rate of change of any of the reactants or products could be used to measure rate. In practice, the rate of use of oxygen is used in most studies. Since carbon dioxide is produced in an equal volume to that of oxygen used, no change in volume would be detected in any apparatus. This is overcome by having a substance in the apparatus which absorbs CO_2.

A simple respirometer

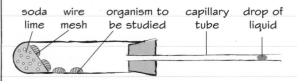

soda lime   wire mesh   organism to be studied   capillary tube   drop of liquid

The drop of liquid will move from right to left as oxygen is used. But there is no system to reset the liquid drop, so no possibility of easily repeating the investigation. There is also no control, or method of controlling temperature.

A more complex respirometer

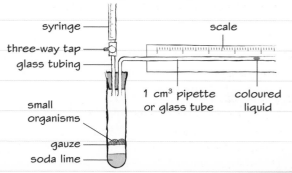

syringe

three-way tap

glass tubing

scale

small organisms

1 cm³ pipette or glass tube   coloured liquid

gauze

soda lime

The syringe and three-way tap allow the return of the coloured liquid, so replicate measurements can be carried out. The test tube can be placed in a thermostatically controlled water bath to control temperature.

Respiration rate

Since respiration is a series of many steps each catalysed by enzymes anything that affects enzymes will affect respiration rate (for reminder of aerobic respiration see page 103). So pH, temperature and enzyme and substrate concentration should all be considered.

Now try this

The table shows the results obtained in an experiment to investigate respiration rate in insects and germinating seeds.

Organism	Distance moved by liquid in 15-minute intervals / nm				Mean rate of respiration / mm min⁻¹
germinating seeds	7	6	5	6	0.4
insects	12	11	13	12	

Calculate the mean and standard deviation for insects. **(4 marks)**

Maths skills The formula to calculate standard deviation (s) is: $s = \sqrt{\dfrac{\sum(x - \overline{x})^2}{n - 1}}$

Worked example

*Explain how you could modify the procedure in the box 'A more complex respirometer' to investigate the effect of temperature on the respiration rate in a small organism. **(9 marks)**

The experiment should be carried out at a range of temperatures not less than 1 °C, or more than 40 °C. Within this range five temperatures should be chosen. For example, 5, 15, 25, 35 and 40 °C. Rate is measured by seeing how far the coloured liquid moves in a known time.

This should be repeated at each temperature so that a mean and standard deviation can be calculated. Placing the test tube in a thermostatically controlled water bath can vary temperature. Other factors include number and size of small organisms and light level, which may affect behaviour if the organism is an animal.

Marks will be awarded for your ability to structure your answer logically, showing how the points that you make are related or follow on from each other where appropriate. Aim to include scientific terminology and biological evidence.

Control of the heartbeat

The cardiac cycle is covered on page 4. Here you look at how that cycle is initiated and controlled.

Cardiac muscle is myogenic

- Heart muscle is **myogenic** because it can beat (contract) without any input from the nervous system.
- Small changes in the electrical charge of cardiac muscle cells cause it to contract.

- The cells are polarised when their outside is negatively charged.
- When the charge is reversed, they are depolarised.
- A wave of depolarisation spreads from cell to cell and causes them to contract.

Control of the heartbeat

1 A heartbeat starts in the sinoatrial node (SAN or pacemaker). This sends electrical impulses across the walls of the atria and they start to contract. A region of non-conducting tissue between the atria and the ventricles prevents the impulse from spreading directly to the ventricles.

2 These impulses reach the atrioventricular node (AVN) and are passed, via the **bundle of His**, to the ventricles.

3 The bundle of His splits into two branches and carries the impulse into the **Purkyne fibres**, which carry the impulse down the ventricles.

4 This impulse causes ventricular contraction from the heart apex upwards, squeezing blood out of the heart.

The ventricles contract after the atria due to the lag in transmission of the impulse via the AVN and the bundle of His.

sinoatrial node (SAN)

atrioventricular node (AVN)

non-conducting layer in heart wall between atria and ventricles

LA

bundle of His

RA

LV

RV

Purkyne fibres   apex of heart

Electrocardiogram

The electrical changes in the heart can be measured and presented as an **electrocardiogram** (ECG).

the QRS complex is the time of ventricular systole

QRS complex   1 second

R

P wave   T wave

Q S

ST segment

PR interval

Electrical changes in a normal heart during one cardiac cycle

the P wave is the time of atrial systole

the T wave is caused by repolarisation of the ventricles during diastole

If disease disrupts the heart's normal conduction pathways there is a disruption of the expected ECG pattern. This can be used for diagnosis of cardiovascular disease.

Examples of abnormal ECGs

1

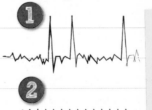

2

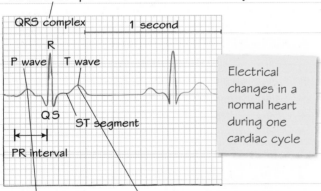

Ventricular tachycardia occurs when electrical impulses arise in the ventricles rather than in the atrium. The ventricles start beating at an abnormally fast, regular rate, and the heart does not work as efficiently.

Atrial fibrillation is when abnormal electrical impulses start firing from sites in the atria. The SAN can no longer control the rhythm of the heart, so the atria contract randomly and too quickly. The heart muscle is unable to relax properly between contractions, reducing the efficiency of the heart.

Worked example

Calculate the heart rate in beats per minute for the person with the normal ECG above. **(2 marks)**

23 squares = 1 beat, 25 squares = 1 second

so 1 beat every $\frac{23}{25}$ sec = 0.92 sec,

so in 1 minute there will be $\frac{60}{0.92}$ beats = 65.2 beats per minute

Now try this

Explain how the electrical activity of the heart ensures that the ventricles begin contracting from the apex of the heart. **(3 marks)**

Cardiac output and ventilation rate

The work of the heart and lungs are linked and coordinated to get oxygen to tissues and take carbon dioxide away from them effectively according to activity level.

 Maths skills ## Cardiac output

Blood is pumped around the body to supply O_2 to and remove CO_2 from respiring tissues. How much is pumped in a minute (cardiac output) depends on two factors:

 how quickly the heart is beating (**heart rate**)

 the volume of blood leaving the left ventricle with each beat (**stroke volume**)

cardiac output (dm³ min⁻¹) = heart rate (min⁻¹) × stroke volume (dm³)

Control of cardiac output

The heart rate can be affected by hormones (for example adrenaline) and nervous control.

The **cardiovascular control centre** in the medulla oblongata of the brain controls the sinoatrial node via nerves (see page 109 for a reminder of how the heartbeat is controlled).

- The **sympathetic nerve** speeds up the heart rate in response to:

  1 fall in pH in the blood due to increased CO_2

  2 lactic acid levels rising

  3 increases in temperature

  4 mechanical activity in joints.

- The **vagus nerve** (parasympathetic) slows down the heart rate as:

  1 the demand for O_2 reduces

  2 the need for removal of CO_2 decreases.

Control of ventilation rate

The rate at which someone breathes is called the **ventilation rate**. You can read how to measure it on page 111.

The **ventilation centre** in the medulla oblongata controls the rate and depth of breathing in response to impulses from chemoreceptors in the medulla and arteries which detect the pH and concentration of CO_2 in the blood. Impulses are sent from the ventilation centre to stimulate the muscles involved in breathing. A small increase in CO_2 concentration causes a large increase in ventilation rate.

The ventilation rate also increases in response to impulses from the motor cortex and from stretch receptors in tendons and muscles involved in movement.

Humans also have voluntary control over breathing

Worked example

The graph shows ECGs of a person with a normal heart and one with tachycardia (a faster than normal heart rate). Analyse the data to explain the effects tachycardia could have on cardiac output.   **(5 marks)**

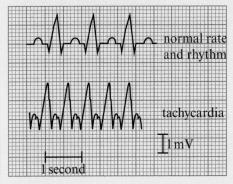

Cardiac output could increase as heart rate increases. The tachycardic person has a heart rate of about 120 min⁻¹, in the normal person it is about 60. On this basis you would expect cardiac output to double. However, we have no information about stroke volume and thus this increase would apply only if the ventricles fill normally at each beat. Cardiac output could decrease if there was insufficient time to fill the ventricles between contractions.

Now try this

Describe the changes in the heart that bring about an increase in cardiac output.   **(4 marks)**

Spirometry

Practical skills In **Core practical 17** you investigate the effects of exercise on tidal volume, breathing rate, respiratory minute ventilation and oxygen consumption using a spirometer.

The spirometer

A person using a **spirometer** breathes in and out of an airtight chamber, causing it to move up and down.

These movements can be recorded on a revolving drum. Alternatively, the moving lid can be attached to a strain gauge that can output to a computer and the movements displayed on a screen.

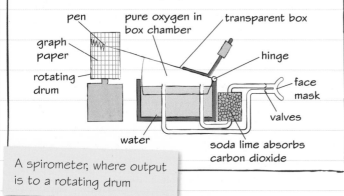

A spirometer, where output is to a rotating drum

The terminology

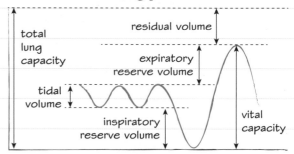

- **tidal volume** is the normal breathing volume, about $0.5\,dm^3$

- **inspiratory and expiratory reserve volume** are the maximum amount you can breathe in and out

- **residual volume** is the air which is in the lungs after you have fully breathed out

- **vital capacity** is the total volume from fully breathed in to fully breathed out.

Trace 1 – without soda lime

Since most respiration is of carbohydrate, the volume of carbon dioxide produced equals the volume of oxygen used. This means that if the carbon dioxide is not absorbed by soda lime the trace will remain horizontal.

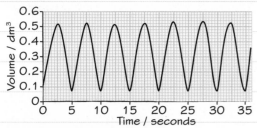

Trace 2 – with soda lime

The soda lime absorbs the carbon dioxide so the subject's use of oxygen causes the trace to slope downwards from left to right.

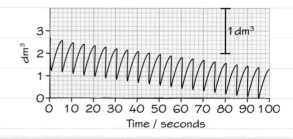

Worked example

Using the traces above calculate tidal volume, breathing rate, respiratory minute ventilation (amount of gas inhaled or exhaled per minute) and oxygen consumption of the subject both at rest and during exercise.

(6 marks)

At rest tidal volume is just about 6 small squares = $0.6\,dm^3$

Breathing rate at rest is 3 breaths in 15 seconds, which gives a rate of 13 breaths min^{-1}

Respiratory minute ventilation at rest = $0.6 \times 12 = 7.2\,dm^3$

Maths skills Tidal volumes can be measured directly from the graph. **Breathing rates** involve counting the number of peaks or troughs over a known time. **Respiratory minute ventilation** is given by:

average tidal volume over 1 minute $\times$ number of breaths over 1 minute

Oxygen consumption is derived from the slope of the trace.

Now try this

The data show the increase in minute ventilation and oxygen uptake in subjects whose heart rate was maintained constant with a pacemaker. Analyse this information to deduce the relative importance of changes in heart rate and minute ventilation of oxygen uptake. **(3 marks)**

Heart rate / beats per minute	Increase in minute ventilation / $dm^3\ min^{-1}$	Increase in oxygen uptake by the blood / $cm^3\ min^{-1}$
50	4.3	87
100	3.9	190

Fast and slow twitch muscle

On page 102 you learned some basics about the structure of muscle fibres, but muscles are not all the same. Mammals have fast and slow twitch fibres, which have different structures and roles.

Muscle fibres (cells) in a muscle

bundle of muscle fibres

myofibril

one muscle fibre (cell)

sarcomere

Diagram showing how muscles are made up from bundles of fibres (cells)

Slow twitch fibres

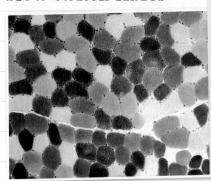

A microscopic view of muscle fibres end on: some are darker than others because they are storing oxygen in **myoglobin** – these are slow twitch fibres

Myoglobin

Myoglobin is a protein similar to haemoglobin. It has a high affinity for oxygen and only releases it when the concentration of oxygen in the cell falls very low, acting as an oxygen store within muscle cells.

Fast and slow twitch compared

Slow twitch	Fast twitch
specialised for slower, sustained contraction and can cope with long periods of exercise	specialised to produce rapid, intense contractions in short bursts
many mitochondria – ATP comes from aerobic respiration (electron transport chain)	few mitochondria – ATP comes from anaerobic respiration (glycolysis)
lots of myoglobin (dark red pigment) to store O_2 and lots of capillaries to supply O_2; this gives the muscle a dark colour	little myoglobin and few capillaries; the muscle has a light colour
fatigue resistant	fatigues quickly
low glycogen content	high glycogen content
low levels of creatine phosphate	high levels of creatine phosphate

Worked example

Explain why greater muscle efficiency may be linked to an increase in the percentage of slow twitch muscle fibres. **(3 marks)**

Slow twitch muscles carry out aerobic respiration so glucose is fully oxidised. This means that more ATP is produced than if respiration were anaerobic. It also takes longer for lactate levels to build up in slow twitch muscle. This means the athlete would suffer cramp less quickly.

Make sure you follow the logic in your answer through. It would be easy to just state that these muscles respire aerobically and leave it at that, but there are 3 marks.

Now try this

Slow twitch muscle fibres have
- ☐ **A** less myoglobin than fast twitch fibres
- ☐ **B** more myoglobin than fast twitch fibres
- ☐ **C** no myoglobin
- ☐ **D** the same quantity of myoglobin as fast twitch fibres. **(1 mark)**

Homeostasis

Homeostasis is the control of such things as temperature, pH and water potential inside the body.

Dynamic equilibrium

Homeostasis is concerned with keeping the conditions for cells within **narrow limits** as close to optimum as possible. Many external factors are very variable so complex organisms have a variety of mechanisms to prevent any changes in the environment from affecting cells. This is known as maintaining a **dynamic equilibrium**.

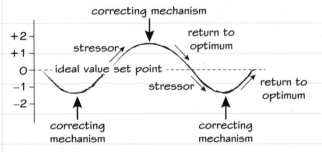

The importance of homeostasis

Stable conditions are essential in order to allow the biochemistry of cells to work efficiently. Proteins, including enzymes, are particularly sensitive to changes. **Intracellular enzymes** (as opposed to extracellular) are extremely fragile.

pH affects molecular structure
- hydrogen bonds disrupted
- charge distribution on active sites changed

temperature affects proteins
- rate of reaction of enzymes
- stability of protein shape

water potential is most important in animals
- affects water content of cells
- too high will cause swelling and bursting
- too low will cause shrinkage of cells.

Control mechanisms

Control of most variables follows the same basic pattern but with differences in the receptors, effectors and means of communication.

Changes are detected and effectors are stimulated to respond to oppose the change.

Once the set point has been restored, the sensory receptors detect the change back to normal and communicate with the effectors to halt the response. This is **negative feedback**.

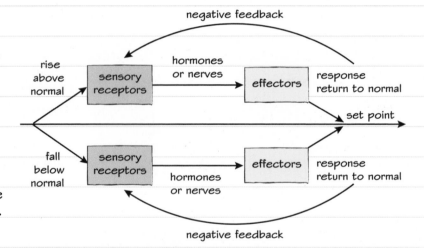

Positive feedback

Positive feedback is rare as it is destabilising. Changes are detected and stimulate responses, which cause further change. An example of positive feedback occurs in the formation of action potentials in nerves where a small increase in membrane permeability stimulates further increases.

Worked example

Explain why negative feedback is more commonly found in control systems than positive feedback. **(2 marks)**

Negative feedback opposes change and maintains stable conditions. Positive feedback amplifies change and makes conditions unstable. Most body systems require stable conditions so negative feedback is most common.

Now try this

Give reasons why enzymes can be affected when homeostatic mechanisms are unable to maintain stable conditions.
(4 marks)

Thermoregulation

Mammals have a very sophisticated temperature regulation system. The body responds to exercise by increasing cardiac output and ventilation rate releasing a lot of heat energy which causes a rise in body temperature.

Thermoregulation

- The control of core body temperature through negative feedback is called **thermoregulation**.
- **Thermoreceptors** in the skin detect changes in temperature.
- Also, thermoreceptors in the **hypothalamus** in the brain detect changes in the **core blood temperature**.
- If a rise in temperature is detected above the normal value, the **heat loss centre** in the hypothalamus will stimulate effectors to increase heat loss from the body – usually through the skin.
- If there is a fall in temperature, the **heat gain centre** takes over.

Heat loss centre	
Stimulates:	• sweat glands to secrete sweat.
Inhibits:	• contraction of arterioles in skin (dilates capillaries in skin) • hair erector muscles (relax – hairs lie flat) • liver (reduces metabolic rate) • skeletal muscles (relax – no shivering).

Heat gain centre	
Stimulates:	• arterioles in the skin to constrict • hair erector muscles to contract • liver to raise metabolic rate • skeletal muscles to contract in shivering.
Inhibits:	• sweat glands.

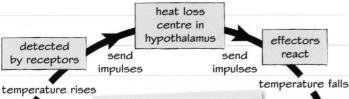

detected by receptors — *temperature rises* — send impulses → heat loss centre in hypothalamus — send impulses → effectors react — *temperature falls*

set point (norm)

The role of the heat loss and the heat gain centres in the hypothalamus of a mammalian brain

set point (norm) — *temperature falls* → detected by receptors — send impulses → heat gain centre in hypothalamus — send impulses → effectors react — *temperature rises* → set point (norm)

Two marathon runners, A and B, had their core temperatures recorded during a race. The graph below shows the results.

During this race, runner A lost 3.02 kg of water and runner B lost 2.43 kg of water.

Analyse this information to explain the change in core temperature of runner A after 120 minutes. **(4 marks)**

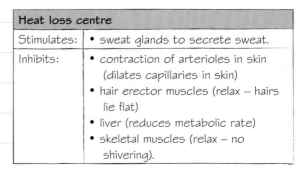

'Analyse …. to **explain**' questions are very common. You must do something with the information. Here it is quite simply to note that the 'change' referred to in the question is an increase. You then go on to use your analysis to explain what you have found.

Runner A's body temperature goes up by 2 °C. Runner A is probably suffering from dehydration and is no longer sweating. Because of this, cooling mechanisms are failing and heat production is still greater than heat loss.

Graph: Core temperature / °C (y-axis 36 to 42) vs Time start since of race / min (x-axis 0 to 160), showing Runner A and Runner B.

Now try this

During a chase of a mammalian prey animal by its mammalian predator, the core body temperatures of both animals rise. Describe how changes in blood circulation help to return their core body temperatures to normal.

(4 marks)

Exercise

Both too little or too much exercise can cause problems.

Possible effects of too little exercise

Too little exercise can lead to:

reduced

- physical endurance, lung capacity, stroke volume and maximum heart rate
- levels of HDL ('good' cholesterol)
- bone density, therefore increased risk of osteoporosis

increased

- resting heart rate, blood pressure, storage of fat in the body and levels of LDL ('bad' cholesterol)

risk of

- coronary heart disease, type II diabetes, some cancers, weight gain and obesity

impaired

- immune response due to lack of natural killer cells.

Possible effects of too much exercise

Overtraining can lead to:

- immune suppression
- increased wear and tear on joints
- chronic fatigue and poor athletic performance
- damage to cartilage in synovial joints, which in turn can lead to inflammation and a form of arthritis
- ligaments being damaged because bursae (fluid-filled sacs) that cushion parts of the joint can become inflamed and tender.

There is some evidence that the number and activity of some cells of the immune system may be decreased while the body recovers after vigorous exercise. It may also be the case that damage to muscles during exercise and the release of hormones such as adrenaline may cause an inflammatory response which could also suppress the immune system.

A note about correlations

Just because two things are observed to happen in tandem, it doesn't mean that one caused the other. A causal link is more likely if you can provide a biological explanation for why one factor could affect the other.

For example, there is a positive correlation between the number of shark attacks and the number of ice creams sold at a beach. There is no biological explanation for this correlation. In contrast, it is thought that there is a causal link between cigarette smoking and lung cancer, because there is a biological explanation about why smoking could cause lung cancer (see page 7).

Analysis and interpretation of data

The analysis of data refers to how the data collected from an experiment are treated and displayed, so that they can be described. In the worked example below, the intensity of exercise is plotted against risk of contracting an upper respiratory tract (URT) infection. The raw data would be the symptoms experienced by people after a period of intense exercise. From this a risk factor would be calculated, and a graph drawn. Interpretation of the data should review whether the analysis is valid and seek to draw conclusions, which would involve making a judgement as to whether the data show a correlation. If they do, a study may go on to try and see if this is causal.

Worked example

In the axes provided, sketch the shape of a graph which would support the conclusion that 'moderate exercise seems to offer some protection against upper respiratory tract (URT) infections.'   **(2 marks)**

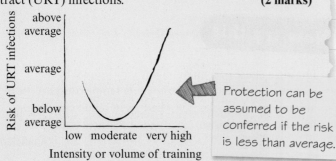

Protection can be assumed to be conferred if the risk is less than average.

Now try this

Two weeks after taking part in a 56 km race, 33% of the runners developed respiratory tract infections. Those who completed the race were three times more likely to develop an infection after the race compared with a control group which did not run. Explain one factor which could contribute to this higher infection rate.   **(3 marks)**

Sports participation and doping

Medical technology can enable those with injuries and disabilities to participate in sport. However, many people deem performance-enhancing substances to be unethical. Are both approaches acceptable?

Surgery and prostheses

- It is possible for surgeons to repair torn cruciate ligaments in the knee using fibre optics to perform **keyhole surgery**. Only a small incision is needed so there is less bleeding and damage to the joint.

- A **prosthesis** is an artificial body part designed to regain some degree of normal function or appearance. The design of prostheses has improved and disabled athletes are now able to compete at a very high level.

- Damaged joints (such as knee joints) can be repaired with small **prosthetic implants** to replace the damaged ends of bones, freeing the patient from a life of pain and restoring full mobility.

Enabling a disabled person to participate in sport is thought to be **ethical**.

Doping in sport

Doping in sport is a vicious circle. Since some athletes use illegal performance-enhancing substances it means that others may follow suit because they don't want to be at a disadvantage. And so it goes on.

Doping in sport could be considered:

- ✗ not acceptable – athletes have a right to compete fairly
- ✓ acceptable because athletes have the right to decide for themselves.

Ethical frameworks

Consider these points when deciding if something, such as using performance-enhancing drugs, is ethical:

- rights and duties
- maximising the amount of good
- making decisions for yourself
- leading a virtuous life.

Hormones in sport

Some drugs such as anabolic steroids are closely related to natural steroid hormones.

They can pass directly through cell membranes and be carried into the nucleus bound to a receptor molecule. These hormone/receptor complexes act as transcription factors. As a result more protein synthesis takes place in the cells.

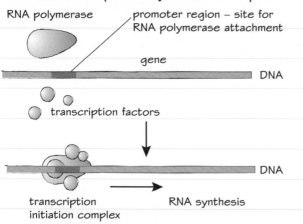

- **Testosterone** increases protein synthesis in muscle cells, increasing the size and strength of the muscle tissue. Peptide hormones bind with receptors on the cell surface membrane. This starts the activation of a transcription factor within the nucleus.

- **Erythropoietin** (EPO) stimulates the production of red blood cells.

Worked example

Amphetamines are similar in chemical structure to certain neurotransmitters. Amphetamines reduce blood flow through the skin during exercise. Justify the view that amphetamines should be made legal.

(3 marks)

Society is generally happy with the idea of enhancement of all kinds. Plastic surgery, slimming pills and implants are used by people to enhance their appearance. Others drink coffee to make them more alert and focused. People may ask why sports participants should be subject to different attitudes or regulations.

Now try this

Performance-enhancing drugs may affect heart activity. Explain one ethical position relating to whether these drugs should be banned. **(3 marks)**

Exam skills

This exam-style question uses knowledge and skills you have already revised. Have a look at page 114.

Worked example

Thermoregulation can cause problems for distance runners. Two marathon runners, A and B, had their core temperatures recorded during a race. The graph shows the results.

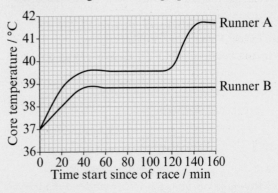

(a) Give reasons for the change in core temperatures in the first 30 minutes of the race. **(2 marks)**

During the first 30 minutes of the race, core body temperature of both runners increases because the increased muscle contraction during running requires increased rate of respiration to provide energy. The heat gained from respiration is greater than heat lost during this period so temperature increases.

(b) Explain the mechanisms that control core temperature between 60 and 100 minutes of the race. **(5 marks)**

- Thermoreceptors in the hypothalamus detect core temperature.
- The heat loss centre will be stimulated by the high core temperature.
- Heat loss mechanisms will be stimulated by the autonomic nervous system under control of the hypothalamus.
- Sweating will cause cooling as water evaporates from the surface of the skin.
- Vasodilation of skin arterioles will increase blood flow to the skin to increase cooling through radiation.
- Blood flow through skin capillaries will increase as sphincter muscles direct blood flow.

(c) Runner A lost over 3 kg of water during the race.

  (i) Deduce reasons for the change in temperature after 120 minutes. **(2 marks)**

Not enough water available to produce sweat. Heat production will exceed heat loss so core body temperature increases.

  (ii) Explain the effect of this water loss on the production of urine in the kidneys. **(3 marks)**

Lower water potential of plasma stimulates ADH release from the pituitary. Collecting ducts made permeable to water and water is reabsorbed into the blood. Urine is concentrated and low in volume.

The graph shows two main things of interest:

- periods of change (both runners in the first 40 minutes and runner A after 120 minutes)
- constant periods (both runners between 40 and about 120 minutes).

Also, be careful to stick to core temperature as given in the stem of the question.

The command words **give reasons** mean that you should explain information given to you – in this case the increases in core temperature at the start of the race. Be careful to make the answer specific to the data – you're not being asked for a complete description of thermoregulation.

You need to give detail of how a constant temperature is maintained (it doesn't matter that the actual temperature is above normal). Be sure to limit your answer to mechanisms that promote heat loss – mechanisms that prevent heat production aren't really relevant when you're exercising. Be precise when discussing blood flow – the blood is flowing in vessels, not just 'in the skin'.

Deduce means you need to work something out – here it is the connection between water loss and inability to sweat.

In part (ii) you need to be selective once more, keeping the answer relevant to urine production in the context of lack of water.

Mammalian nervous system

The mammalian nervous system is composed of the central and peripheral nervous systems. The peripheral nervous system is divided into autonomic and somatic systems. The autonomic nervous system is divided into sympathetic and parasympathetic systems, which act antagonistically.

Structure of the nervous system

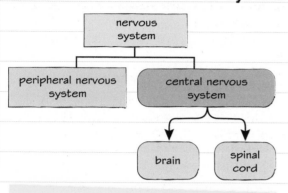

The nervous system of a mammal is divided into the **central nervous system** (CNS) and the **peripheral system**. The CNS is made up of the brain and the spinal cord.

The peripheral nervous system

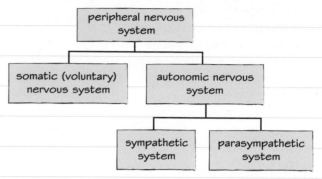

The **somatic nervous system** is concerned with conscious or voluntary activity. The **autonomic nervous system** is concerned with controlling involuntary actions, such as heart rate.

The autonomic system has antagonistic branches that act in opposite ways. For example, the sympathetic system is responsible for increasing heart rate, whereas the parasympathetic system slows heart rate.

Types of neurone

The mammalian nervous system is made up of cells called **neurones**. There are several types of neurone. The three most common types are shown here.

sensory neurone

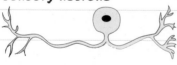

Connects sensory receptors to the central nervous systems

relay neurone

Found in the central nervous system

motor neurone

Communicates from central nervous system to effectors

Structure of a neurone

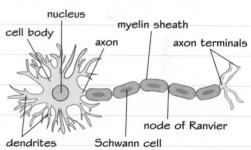

This shows a motor neurone but the parts are similar in sensory neurones. The Schwann cells wrap round the axon to insulate it.

Worked example

Use the diagrams on this page.

Compare and contrast the structure of motor and sensory neurones. **(4 marks)**

Both have a long nerve fibre.

In a motor neurone the nerve fibre is an axon, in a sensory neurone it is a dendron.

Motor neurones have more dendrons and dendrites on the cell body.

Sensory fibres have their cell body displaced to the side of the fibre, motor neurones have the cell body at the beginning of the nerve fibre.

Now try this

Explain how the parasympathetic and sympathetic branches of the autonomic nervous system work antagonistically. **(3 marks)**

Stimulus and response

Throughout the nervous system, effectors such as muscles respond to a stimulus. This can be illustrated by changes in size of the pupil of the eye in response to changes in light intensity.

Stimulus and response system (a reflex)

- Animal nervous systems are fast-acting communication systems containing nerve cells (**neurones**) that carry information in the form of nerve impulses.

- In mammals, sensory neurones carry impulses from receptors to a central nervous system (CNS) consisting of the brain and spinal cord.

- The CNS (containing relay neurones) processes information from many sources and then sends out impulses via motor neurones to effector organs (mainly muscles and glands).

The pupil reflex

- The iris contains pairs of antagonistic muscles (**radial and circular muscles**) that control its size.

- These muscles are under the influence of the autonomic (involuntary) nervous system.

- In high light intensity, photoreceptors in the retina cause nerve impulses to pass at high frequency along the optic nerve to nerve cells in the brain. These then send impulses along parasympathetic motor neurones to the circular muscles of the iris. The muscles contract, reducing the diameter of the pupil so that less light can enter the eye.

- In low light intensity, impulses at low frequency reach the coordinating centre in the brain so impulses are sent down sympathetic motor neurones to the radial muscles of the iris instead. This causes the radial muscles to contract and the pupil becomes dilated, allowing more light to reach the retina.

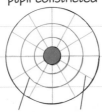

pupil constricted   pupil dilated

circular   radial   circular   radial
muscles   muscles   muscles   muscles
contract   relax   relax   contract

How the muscles of the iris act to control the amount of light entering the eye

---

The photo on the left shows a cat's eyes in bright light. The photo on the right shows a cat's eyes in dim light.

Explain how the nervous system has controlled the change in the cat's pupils in dim light.   **(4 marks)**

In the dim light, less light hits photoreceptors on the retina. This means that fewer impulses pass to the brain along the sensory neurone. The brain sends impulses along the sympathetic motor neurones causing radial muscles to contract, the circular muscles relax and the pupil dilates.

Although this question is basically testing knowledge, you do have to think it through so you get everything the right way round. Try writing an answer explaining the change in pupil diameter when the light is switched on.

Name the effector and the receptor in the pupil reflex.   **(2 marks)**

The resting potential

The **resting potential** describes the state of nerve fibres when they are not conducting an impulse. It is a dynamic equilibrium that results from an imbalance in concentrations of Na^+ and K^+ ions and differences in permeability of the cell membrane to ions.

Potential difference (PD)

The **potential difference** measures the difference in charge across a membrane. It is always a comparison, in this case comparing inside the cell cytoplasm with outside of the cell.

Sodium potassium pumps

Sodium potassium pumps are proteins in the cell membrane. The role of the pumps is to maintain diffusion gradients for K^+ and Na^+ across the membrane. They work in the background but they are not essential for individual impulses.

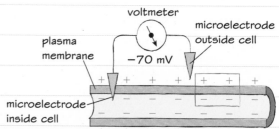

Potential difference can be measured using tiny electrodes. Here, the inside of the nerve fibre is 70 mV more negative than outside. The resting potential of nerve cells is usually around -70 mV.

Channel proteins

The action of the Na^+/K^+ pump results in an unequal concentration of ions:

* higher Na^+ outside nerve fibre
* higher K^+ inside nerve fibre

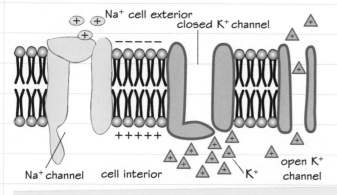

At rest, K^+ channels are proteins allowing movement of ions across the membrane. These are always open.

Resting potential equilibrium

The resting potential results from two opposing forces on K^+ ions.

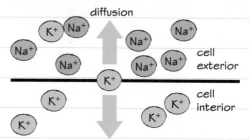

Balance of movement of K^+ causes the resting potential

* large diffusion gradient for K^+ to leave cell
* K^+ diffuses out of cell taking positive charge
* inside is left more negative than outside
* negative charge in cell stops diffusion of K^+
* balance of opposing forces at -70 mV
* there is very little movement of Na^+ back into the cell because the membrane is impermeable to it.

Worked example

The concentration of some ions was measured within an axon and in the fluid surrounding the axon. The results were as follows.

	Inside cell / mmol	Outside cell / mmol
Na^+	10	120
K^+	140	4

The potential difference across the membrane was found to be more negative than -70 mV.

Deduce the effect of adding KCl to the external solution on the potential difference. **(4 marks)**

KCl would add K^+ and Cl^- to the external solution.

This would not affect the potential difference as the charges added balance each other.

The concentration gradient for K^+ would be reduced.

Fewer K^+ would diffuse out as the potential difference needed to oppose the diffusion gradient would be lower so the potential difference would be less than -70 mV.

Now try this

Describe the role of the cell membrane of an axon in maintaining the resting potential. **(3 marks)**

Action potential

An **action potential** (nerve impulse) is a change in the potential difference (PD) at a point on the membrane of a nerve cell. Action potentials allow information to be passed along nerve fibres.

Changes in polarity

The action potential is triggered by the depolarisation of the nearby membrane changing the PD to the threshold potential.

As the result of an influx of sodium and potassium ions, the polarity of the membrane is reversed so that inside the cell changes from negative ($-70\,mV$) to positive ($+40\,mV$).

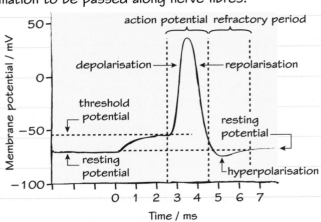

Changes in membrane potential of a nerve before, during and after an action potential

Changes in membrane permeability

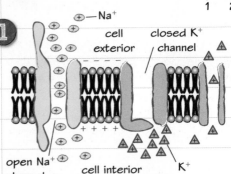

1 depolarisation

2 repolarisation

3 recovery

1
open Na⁺ channel
cell exterior — Na⁺
closed K⁺ channel
cell interior K⁺

- Na⁺ gates open
- Na⁺ diffuses into the cell carrying positive charge
- Na⁺ gates close.

2
Na⁺ cell exterior
open K⁺ channel
closed Na⁺ channel   cell interior   K⁺

- K⁺ gates open
- K⁺ diffuses out of the cell taking positive charge with it
- K⁺ gates close.

3
cell exterior
closed K⁺ channel

- K⁺ moves back into cell attracted by negative charge because they are hyperpolarised (more negative than $-70\ mV$)
- resting potential equilibrium restored.

Worked example

The diagram shows changes of potential difference during an action potential.

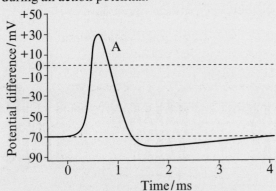

Explain how the resting potential is restored after $1.5\,ms$. **(3 marks)**

The resting potential is restored by closing of the voltage-dependent K⁺ channels and potassium ion diffusion into the axon through non-voltage gated channels, which causes the PD to rise.

Don't be tempted to mention the refractory period here as that's only relevant to the movement of the impulse along the fibre.

Also, although the Na⁺/K⁺ pumps maintain the diffusion gradients in the long term they're not important to a single impulse.

Now try this

Describe and explain the movement of ions at point A in the diagram shown in the worked example. **(4 marks)**

There are two different forces causing ions to move at this point.

Propagation of an action potential

An action potential moves along a nerve fibre as potential differences change, affecting membrane proteins. The speed of transmission along myelinated axons is greater than along non-myelinated axons as a result of **saltatory conduction**.

Propagation of the action potential

- Na^+ entering during the action potential flows along the inside of the nerve fibre.
- The additional positive charge inside the membrane will reduce the potential difference to the threshold, which opens voltage-dependent sodium gates in the membrane (shown as -40 mV here).

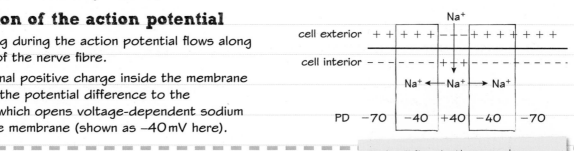

Refractory period

A new action potential cannot be generated in the same section of membrane for about five milliseconds. This is known as the **refractory period**, which makes sure that the impulse travels in one direction along a nerve fibre.

Na^+ will flow both ways along the membrane but the refractory period, from where the impulse has come, prevents it returning. The flow of ions along the nerve fibre is known as a **local circuit**.

Myelinated nerve fibres

For nerve cells without a myelin sheath, the speed that an impulse travels depends on its cross sectional area. Nerves with a bigger cross-sectional area work faster.

Mammals need fast impulses but their nerve fibres are small so they use a myelin sheath to speed conduction by limiting depolarisation to the **nodes of Ranvier**, which are gaps between insulating **Schwann cells**.

Saltatory (jumping) conduction

The only region of a myelinated nerve fibre that can depolarise is at the nodes of Ranvier.

This means that the local circuits cover a much longer distance than they would without myelin, and the impulse 'jumps' from one node to the next.

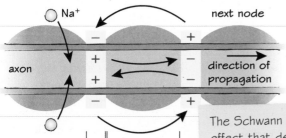

The Schwann cells have the effect that depolarisation of a node causes depolarisation of the next node, resulting in much faster conduction along the nerve fibre.

Worked example

The diameter and rate of conduction of some nerve fibres was measured. The results are shown in the table.

Nerve fibre	Diameter / µm	Speed / m s⁻¹
mouse	18	96
squid	80	25
earthworm	2	5

(a) Calculate the difference in cross-sectional area of the squid and earthworm nerve fibres. **(3 marks)**

The cross-section of a fibre is a circle, so using πr^2
mouse is $3.14 \times 9^2 = 254$ µm²
squid is $3.14 \times 40^2 = 5024$ µm²
difference is 4770 µm²

(b) Discuss the differences in speed of conduction of the three nerve fibres. **(5 marks)**

The speed of conduction is faster in fibres with larger diameters. The squid fibre is much bigger in diameter than the earthworm, so has much faster conduction. Mouse fibres are myelinated and myelination speeds up conduction due to saltatory conduction. So, the mouse fibres are likely to be the fastest despite having a smaller diameter than squid fibres.

 Maths skills The area of a circle can be calculated using the formula: $A = \pi r^2$, where r is the radius of the circle. Use either the π button on your calculator or the rounded value 3.14.

Now try this

Explain the significance of the refractory period to nervous conduction. **(4 marks)**

Synapses

Action potentials cannot pass between nerve cells across the synapse, so transmitter substances are used.

Structure of a synapse

A synapse is the junction between two neurones. The **presynaptic membrane** (before the gap) allows the release of chemicals (**neurotransmitters**) when impulses are arriving to stimulate impulses in the cell after the gap (**postsynaptic cell**).

A typical synapse has a presynaptic knob containing vesicles filled with neurotransmitter. Receptor molecules on the postsynaptic membrane bind with neurotransmitter.

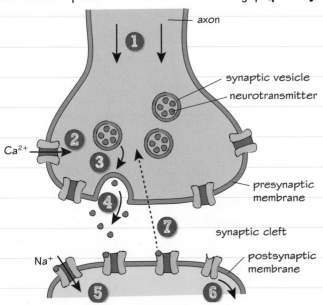

1 An action potential arrives.

2 The membrane depolarises. Calcium ion channels open. Calcium ions enter the neurone.

3 Calcium ions cause synaptic vesicles containing neurotransmitter to fuse with presynaptic membrane.

4 Neurotransmitter is released into the synaptic cleft.

5 Neurotransmitter binds with receptors on the postsynaptic membrane. Sodium channels open. Sodium ions flow through the channels.

6 The membrane depolarises and initiates an action potential.

7 When released, the neurotransmitter will be taken up across the presynaptic membrane (whole or after being broken down), or it can diffuse away and be broken down.

Acetylcholine (ACh)

- found in all nerves of the voluntary and the parasympathetic autonomic system (see page 113)
- the enzyme which breaks it down in a synapse is acetylcholinesterase.

Postsynaptic potentials

Binding of neurotransmitters to receptors at the postsynaptic membrane will affect the potential difference of the postsynaptic membrane, depending on which type of receptor binds the neurotransmitter.

- a receptor with an **excitatory synapse** sends **excitatory postsynaptic potentials** (EPSP)
  - opening of sodium ion channels reduces the potential difference to closer to the threshold potential
- a receptor with an **inhibitory synapse** sends **inhibitory postsynaptic potentials** (IPSP)
  - postsynaptic membrane is hyperpolarised moving further from the threshold.

Many nerve pathways have a mixture of both types of receptor. Once enough EPSPs have arrived a new action potential is triggered in the postsynaptic neurone.

Always talk about a high enough frequency of impulses arriving at a synapse – one impulse does not trigger an action potential in the postsynaptic membrane. It takes a number of impulses to have an effect.

Worked example

Serotonin is a neurotransmitter found in synapses in the brain. An excess of serotonin can cause anxiety.

serotonin   MDMA

A drug called MDMA can also cause similar feelings of anxiety. The diagrams represent the shapes of serotonin and MDMA molecules. Deduce how MDMA causes anxiety. **(4 marks)**

The two molecules have similar shapes.

This could allow both to bind to postsynaptic receptors.

Binding will cause postsynaptic potentials to develop.

The nerve pathways causing anxiety will be stimulated by MDMA in the same way as serotonin giving the effects of excess serotonin.

Now try this

Explain the difference between excitatory and inhibitory postsynaptic potentials. **(3 marks)**

Vision

Receptors are specialised cells able to detect stimuli. Receptors are often grouped together into sense organs. The eye is an excellent example.

Human photoreceptors

Human eyes have two types of **photoreceptor** cell found in the retina on the back of the eye.

1. **Cones** allow colour vision in bright light, and are clustered in the centre of the retina at the back of the eye.

2. **Rods** only provide black and white vision, but are much more sensitive to light intensity than cones and can work in dim light conditions.

pigment epithelium

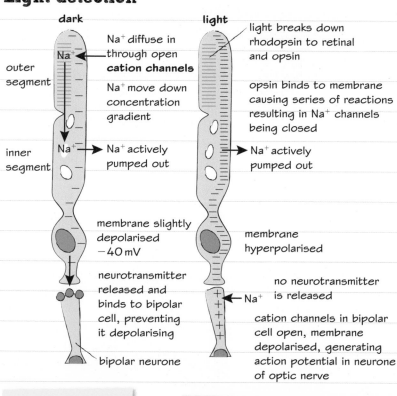

rod   cone

Rods and cones make up the retina

How a rod cell detects light

1 Light energy is absorbed by **rhodopsin** which splits into **retinal** and **opsin**.

2 Without retinal, the opsin binds to the membrane of the outer segment of the cell.

3 This causes cation channels to close. The inner segment continues to pump sodium ions out of the cell and the membrane becomes **hyperpolarised** (more negative).

4 This means that glutamate is not released across the synapse. Glutamate usually inhibits the neurones that connect the rod cells to the neurones in the optic nerve.

5 When there is less inhibition an action potential forms and is transmitted to the brain. The information from the optic nerve is processed by the brain in the visual cortex.

Light detection

dark

outer segment — Na⁺ diffuse in through open **cation channels**

Na⁺ move down concentration gradient

inner segment — Na⁺ actively pumped out

membrane slightly depolarised −40 mV

neurotransmitter released and binds to bipolar cell, preventing it depolarising

bipolar neurone

light

light breaks down rhodopsin to retinal and opsin

opsin binds to membrane causing series of reactions resulting in Na⁺ channels being closed

Na⁺ actively pumped out

membrane hyperpolarised

no neurotransmitter is released

cation channels in bipolar cell open, membrane depolarised, generating action potential in neurone of optic nerve

How a rod changes in the light

ATP is not mentioned in the diagram. Recall its role and apply that to what you see in the diagrams.

Worked example

Groups of three rod cells connect to a single bipolar cell, whereas just one cone cell connects to a bipolar cell. Use this information to explain why you can't see colour well in dim light conditions. **(3 marks)**

Rods are more sensitive than cones but do not detect differences in colour.

Three rod cells converging on a single bipolar cell have a greater chance of affecting the bipolar cell in low light levels than a single cone cell.

Vision in dim light is based on non-colour sensitive rods which are not sensitive to colour.

Now try this

Using the information in the diagrams above, describe the role of ATP in the hyperpolarisation of rod cells in the retina. **(2 marks)**

Plant responses

Plants can detect and respond to light direction, duration, intensity and wavelength. If a plant grows towards a stimulus it is said to be a positive tropic response. **Photoperiodism** is the response to light duration.

Photoperiodism

Plants flower and seeds germinate in response to changes in day length. The photoreceptor involved is a blue-green pigment called **phytochrome**.

On absorbing natural (or red) light, phytochrome converts from the inactive form Pr (r stands for red) to the active form Pfr (fr stands for far red). In the dark, Pfr slowly reverts back to Pr because it is relatively unstable or it can change back rapidly into Pr if exposed to far red light. It is thought that the active Pfr may trigger a range of different photoperiodic responses, such as flowering, germination and greening.

How tropisms work

Phototropism is the response to direction of light. **Tropisms** are growth responses in plants where the direction of the growth response is determined by the direction of the external stimulus.

with illumination from all sides, an even distribution of auxins moves down from the shoot tip, and causes elongation of cells across the zone of elongation

zone of elongation

auxins broken down by enzymes

with illumination from one side, auxins move down from the shoot tip towards the shaded side of the shoot; only those cells on the shaded side elongate, and the shoot bends towards the light

Mechanism of phototropism in shoots

The receptor is in the tip and it is thought it might be riboflavin. The effector for the growth response is cell elongation. This happens just below the tip of the shoot and is controlled by the plant growth substance IAA (the first auxin discovered).

This is **positive phototropism**. Growth away from light, shown by some roots, is **negative phototropism**.

Other environmental cues

- **gravitropism** is the response to gravity
- **chemotropism** is the response to chemicals
- **hydrotropism** is the response to water.

What do IAA and Pfr do?

- bind to protein receptors in the target cells
- activate intracellular second messenger signal molecules
- these activate transcription factors, which control the transcription of genes
- the proteins produced bring about metabolic changes that result in many responses, including cell expansion, division and differentiation.

Worked example

A study was carried out to investigate the effect of red light and far red light on sunflower plants. One group of seedlings, A, was grown under a lamp that emitted red light and far red light of the same intensity. Another group, B, was grown in the same way, except that the lamp emitted a lower intensity of red light. The mean dry mass of the flowers produced and the mean length of the plant stems were recorded. The results of the study are shown in the table.

Study	Mean dry mass of the flowers / g		Mean stem length / cm	
	Group A	Group B	Group A	Group B
original	58	45	125	148

Analyse these data to explain the importance of these light conditions for a young seedling in the woodland, where conditions experienced by group B are found.

(3 marks)

There is an increase in stem length in B conditions compared to A.

This will give a taller seedling.

Such a seedling will receive more light and maximise photosynthesis in parts of woodland with less light.

Now try this

Give three similarities between IAA and animal hormones.   (3 marks)

Nervous and hormonal control

In response to stimuli from the environment, the two **coordination systems** (the nervous and the endocrine systems) work to process information and bring about coordination.

The nervous system

Neurones transmit information in the form of nerve impulses. (Look at page 118 to remind yourself of the mammalian nervous system.)

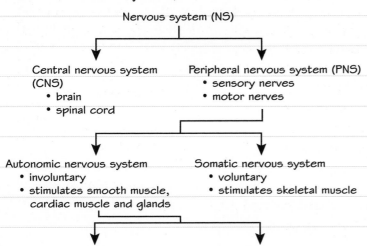

The organisation of the nervous system

The endocrine system

Hormones regulate and coordinate functions of the body that require maintained responses in the **endocrine system**.

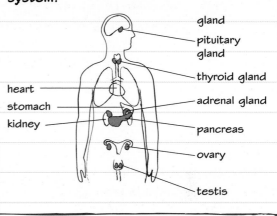

Nervous and endocrine systems compared

Nervous system in mammals	Endocrine system in mammals
electrochemical changes giving an electrical impulse; chemical neurotransmitters used at most synapses	chemical hormones from endocrine glands carried in the blood plasma around the circulatory system
rapid acting	slower acting
usually associated with short-term changes, e.g. muscle contraction	can control long-term responses, e.g. growth and sexual development; some are involved in homeostasis, e.g. control of blood sugar; some can be relatively fast, e.g. effects of adrenaline in response to stress
response is very local and specific such as a muscle cell or gland	response may be widespread, or restricted to specific target cells

With all MCQ questions, you should eliminate the answers you think are incorrect and then go the other way too; it is a 'belt and braces' approach.

Worked example

Which of the following correctly compares nervous coordination with hormonal coordination? **(1 mark)**

A nervous coordination is faster and lasts for a longer time

B nervous coordination is faster and lasts for a shorter time

C nervous coordination is slower and lasts for a longer time

D nervous coordination is slower and lasts for a shorter time

B

Now try this

Describe how **one** physiological function is controlled by both nerves and hormones.

(3 marks)

The human brain

The central nervous system is made up of the spinal cord and the brain. The brain has areas with clearly defined functions.

Brain structure

The brain has distinct regions and particular functions have been identified for each.

> A section through the brain shows the main areas and their main functions.

cerebral hemispheres the cerebrum is associated with advanced mental activity like language, memory, calculation, processing information from the eyes and ears, emotion and controlling all of the voluntary activities of the body

hypothalamus controls thermoregulation

medulla oblongata controls many body processes such as heart rate, breathing and blood pressure

cerebellum for balance and coordinating muscle movements

Imaging techniques for the brain

- **Magnetic resonance imaging** (MRI) scans use a magnetic field and radio waves to make images of soft tissues like the brain. MRI scans can be used to diagnose tumours, strokes, brain injuries and infections, and to track degenerative diseases like Alzheimer's by comparing scans over time.

- **Computed tomography** (CT) (or **computerised axial tomography** (CAT)) scans use thousands of narrow beam X-rays rotated around the patient. They only capture one moment in time and use harmful X-rays.

- **Positron emission tomography** (PET) scans use isotopes with short half-lives, such as carbon-11 in glucose and other molecules. These act as radiotracers which are injected. They are detected on positron emission. Increased blood flow to active areas of the brain show up as a bright spot on the scan. PET is useful in the diagnosis and monitoring of Alzheimer's disease.

Functional magnetic resonance imaging (fMRI)

The effects of drugs and diseases, such as Parkinson's disease, on brain activity can be seen using fMRI. A modified MRI technique detects activity in the brain by following the uptake of oxygen in active brain areas, so it allows you to see the brain in action during live tasks.

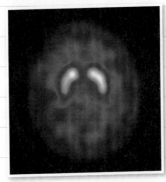 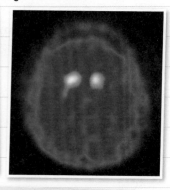

fMRI scans of a healthy brain, a patient with Parkinson's disease without drug treatment and a patient with Parkinson's disease taking drug treatment. The scan shows a horizontal section with the front of the head at the top. The most active areas are white.

Now try this

Compare and contrast the use of computed tomography (CT) with magnetic resonance imaging (MRI) for studying brain structure.   **(4 marks)**

Critical window for development

We are born with a range of innate behaviours (behavioural responses that do not need to be learned) such as crying, grasping and sucking. However, the brain still needs much growth and development after birth through the formation of synapses and the growth of axons.

What happens during the critical window

Critical windows for development are periods of time where it is thought that the nervous system needs specific stimuli in order to develop properly.

Eye deprived of light during critical window	Eye that remains open during the critical window
axons do not pass nerve impulses to cells in the visual cortex	axons pass nerve impulses to cells in the visual cortex
inactive synapses are eliminated	synapses used by active axons are strengthened
eye has no working connection to the visual cortex and is effectively blind, even though the cells of the retina and optic nerve work normally when exposed to light	synapses only present for axons coming from the light-stimulated eye, so the visual cortex can only respond to this eye

Evidence for critical windows

Evidence for critical windows for development has come from:

- medical observations (e.g. children who develop cataracts before the age of 10 days may suffer from permanent visual impairment even if the cataracts are repaired at a later date)
- animal models (e.g. work in monkeys has shown that both light and patterns are needed for full visual capacity development).

Hubel and Wiesel

Kittens and monkeys were used as models to investigate the critical window (or critical period) in visual development because of the similarity of their visual systems to that of humans. They were deprived of light in one eye from birth and after six months were found to be blind, providing evidence for the critical window.

Use of animals for research

Absolutist view of animal rights says humans should never use animals in research.

Relativist view says that humans should minimise harm. The emphasis is on animal welfare, respecting rights to food, water, veterinary treatment and the ability to express normal behaviours.

Utilitarian view says certain animals can be used in medical experiments provided the overall expected benefits are greater than the overall expected harms.

Worked example

*An investigation that used animals was carried out by Hubel and Weisel. These scientists used kittens to investigate brain development. Explain how this work helped to explan human brain development. **(9 marks)**

They investigated development of the visual cortex and found evidence for a critical window for development in this part of the brain. This is a period when neural connections are made in the brain.

The animals were deprived of the stimulus of light into one eye **(monocular deprivation)** at different stages of development and for different lengths of time.

They found that the kittens deprived of light in one eye at 4 weeks after birth were permanently blind in that eye. Monocular deprivation before 3 weeks and after 3 months had no effect.

It was thought that during the critical period (about 4 weeks after birth) connections to cells in the visual cortex from the light-deprived eye had been lost. This meant that the eye that remained open during development became the only route for visual stimuli to reach the visual cortex.

The * indicates that marks will be awarded for your ability to structure your answer logically showing how the points that you make are related, or follow on from each other where appropriate. Aim to include relevant biological evidence and/or scientific facts to support your points.

Now try this

Explain how a scientist might justify the use of animals in experiments.

(3 marks)

Learning

Learning is a process that results in a change in behaviour (or knowledge) as a result of experience. Memories (conscious and subconscious) are formed by changing or making new synapses in the nervous system.

Habituation

Habituation is a very simple type of learning that involves the loss of a response to a repeated stimulus that fails to provide any form of reinforcement (reward or punishment). It allows animals to ignore unimportant stimuli so that they can concentrate on more rewarding or threatening stimuli.

1 with repeated stimulation, Ca^{2+} channels become less responsive so less Ca^{2+} crosses the presynaptic membrane

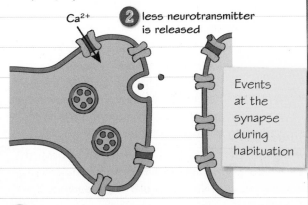

2 less neurotransmitter is released

Events at the synapse during habituation

3 there is less depolarisation of the postsynaptic membrane so no action potential is triggered in the motor neurone

Storage of memories – snails

1 The snail's eye stalk is touched.

2 Impulses pass along a sensory neurone from the eye stalk.

3 The sensory neurone synapses with a motor neurone that connects to the tentacle muscle.

4 Impulses pass along the motor neurone.

5 The eye stalk muscle contracts and the eye stalk is withdrawn.

6 With repeated stimulation the calcium ion channels of the presynaptic neurone become less responsive to the changes in voltage associated with action potentials.

7 Fewer calcium ions enter the presynaptic neurone.

8 Less neurotransmitter is released from the presynaptic neurone.

9 Fewer sodium ion channels are opened in the postsynaptic neurone, so there is less depolarisation of the membrane.

10 An action potential is not generated in the postsynaptic membrane.

11 The eye stalk muscle does not contract and is not withdrawn.

Practical skills — Snails: habituation to a stimulus – Core practical 18

Core practical 18 requires you to investigate habituation to a stimulus.

- The snail is firmly touched between the eye stalks using a cotton bud.
- The stopwatch is simultaneously started.
- The length of time between the touch and the eye stalk being fully emerged is measured.
- Procedure repeated, a total of 15 touches.

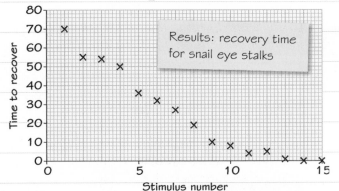

Results: recovery time for snail eye stalks

Worked example

Analyse the information in the graph to explain the effect of stimulus number on the recovery time for snail eye stalks. **(3 marks)**

The time to recover decreases. This is because calcium ion channels become less responsive in sensory neurones so fewer calcium ions are taken up.

This means less neurotransmitter is released so fewer impulses will pass to the muscle.

Now try this

Explain how the learned response shown in the graph may be of benefit to the snail in its natural environment.

(4 marks)

Brain development

Brain development is controlled by both genes (nature) and the environment (nurture).

Nature versus nurture

Nature: Many of our characteristics develop solely under the influence of our genes with little influence from our environment or learning, e.g. blood group.

Nurture: Many characteristics are learned or are heavily influenced by the environment, e.g. how long your hair is.

Most of our characteristics are actually determined by both.

Evidence

Evidence for the relative roles of nature and nurture in brain development come from studies of:

1 newborn babies – they have some innate capacities that suggest genes help to form the brain and some behaviours before the baby is born

2 patients who have suffered from brain damage – some of them show the ability to recover some of their brain function, demonstrating that some neurones have the ability to change

3 experiments such as those carried out by Hubel and Weisel on critical windows for sight (page 129) suggest that external stimulation is important in brain development

4 identical twins – they share the same genes whereas fraternal (non-identical) twins share the same number as any other sibling would; twin studies help to estimate the relative contribution of genes and the environment since differences between identical twins must be due to environmental effects; identical twins raised apart compared to those raised together are very useful; in general, if genes have a strong influence on the development of a characteristic, then the closer the genetic relationship, the stronger the correlation will be between individuals for that trait

5 investigations into the visual perception of groups from different cultural backgrounds – these investigations support the idea that visual cues for depth perception are at least partially learned.

Worked example

In an investigation, each member of a pair of twins were shown a number of human faces and then asked to identify them amongst a group of unfamiliar faces. The agreement in face identification between each pair of twins was recorded.

The results were used to calculate the mean percentage agreement in face identification for the two types of twin. This is shown in the table.

Mean percentage agreement in face identification	
identical twins	non-identical twins
70	29

The scientists concluded that face identification has a genetic component.
Explain how these results support their conclusion.

(4 marks)

In identical twins the agreement is greater by 42%.

This suggests that genes are involved, and non-identical twins have genetic differences.

However, agreement is less than 100% so some other factor is also involved.

Maths skills Note that we do not know the sample size so it would be impossible to reach a conclusion as to whether the difference is **statistically significant** or not. If we did have raw data, the **Chi square test** could be applied to test for statistical significance.

Now try this

In one study of 69 pairs of identical twins, 50 pairs of the twins had been brought up together since birth and 19 pairs of the twins had been brought up apart since birth. The height, body mass and intelligence (IQ) of each twin was measured and the difference between each pair was determined for each characteristic. The table shows the mean differences between the pairs of twins.

Characteristic	Mean difference	
	50 pairs of identical twins brought up together	19 pairs of identical twins brought up apart
height / cm	1.7	1.8
body mass / kg	1.9	4.5
intelligence / IQ	3.1	6.0

Analyse the data to explain the effect of nature and nurture on these three characteristics. **(4 marks)**

Brain chemicals

Parkinson's disease and depression have been shown to be associated with imbalances of important brain chemicals. This knowledge is allowing the development of drugs for the treatment of these conditions.

Dopamine and Parkinson's disease

The symptoms of Parkinson's disease are muscle tremors (shakes); stiffness of muscles and slowness of movement; poor balance and walking problems; difficulties with speech and breathing; depression.

Parkinson's disease is associated with the death of a group of dopamine-secreting neurones in the brain (*substantia nigra*). This results in the reduction of dopamine levels in the brain. Dopamine is a neurotransmitter that is active in neurones in the frontal cortex, brain stem and spinal cord. It is associated with the control of movement and emotional responses.

Treatment of Parkinson's disease

A variety of treatments are available for Parkinson's disease, most of which aim to increase the concentration of dopamine in the brain. Dopamine cannot move into the brain from the bloodstream, but the molecule that is used to make dopamine can. This molecule is called L-dopa and can be turned into dopamine to help control the symptoms.

Serotonin and depression

Serotonin is a neurotransmitter linked to feelings of reward and pleasure. A lack of serotonin is linked to clinical depression (prolonged feelings of sadness, anxiety, hopelessness, loss of interest, restlessness, insomnia). Ecstasy (MDMA) works by preventing the reuptake of serotonin. The effect is the maintenance of a high concentration of serotonin in the synapse, which brings about the mood changes in the users of the drug (see also page 99).

How synapses are affected by drugs

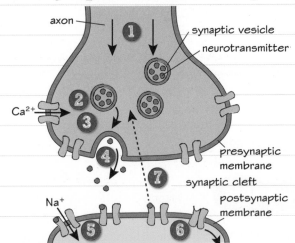

Many drugs affect the nervous system by interfering with the normal functioning of a synapse. Some drugs:

1 affect the synthesis or storage of the neurotransmitter.

2 may affect the release of the neurotransmitter from the presynaptic membrane.

3 may affect the interaction between the neurotransmitter and the receptors on the postsynaptic membrane.

4 may be stimulatory by binding to the receptors and opening the sodium ion channels.

5 may be inhibitory, blocking the receptors on the postsynaptic membranes and preventing the neurotransmitters binding.

6 prevent the reuptake of the neurotransmitter back into the presynaptic membrane.

7 may inhibit the enzymes involved in breaking down the neurotransmitter in the synaptic cleft.

Worked example

A serotonin selective reabsorption inhibitor (SSRI), for example Prozac, may be given to patients to reduce depression. Explain how this helps to reduce depression. **(3 marks)**

Serotonin is not reabsorbed because SSRi binds to reuptake proteins. This means there is a high level of serotonin. Consequently, the increased level of serotinin in the synapses continues to bind to receptors in postsynaptic membranes, increasing the feelings of reward and pleasure.

Now try this

Patients with Parkinson's disease have little of the neurotransmitter dopamine in the motor cortex of their brains. Explain how 'dopamine agonists' might be a useful treatment for Parkinson's disease.

(3 marks)

Dopamine agonists have a similar shape to dopamine.

HGP – the Human Genome Project

A genome is all the DNA (or genes) of an organism. The **Human Genome Project** (HGP) was a multinational project that determined the base sequence of the human genome.

HGP: the key findings

1. Many new genes have been identified, including some of those genes responsible for inherited diseases.

2. In addition new drug targets (specific molecules that drugs interact with to have their effects, e.g. enzymes) have been identified.

3. Information about a patient's genome may help doctors to prescribe the correct drug at the correct dose.

4. Data from the Human Genome Project may also allow some diseases to be prevented.

5. If you understand what genes you carry you may understand what disease you are likely to be at risk from.

The ethics of HGP

Like many scientific advances, HGP raises many questions.

- Who owns the information? Some groups have applied for patents on genetic sequences to try to gain ownership.
- Who is entitled to know the information about your genome if it is sequenced?
- Should insurance companies have access to the information?
- Will genetic screening lead to eugenics (the genetic selection of humans) and designer babies?
- Who will pay for the development of the new therapies and drugs? Many possible highly specialised treatments may be very expensive and will only be suitable for a few people.

In this question it is clear that you need to give both sides of the argument. In others it may not be so obvious, but always think about whether or not you should.

Worked example

Explain how compulsory genetic screening of everyone might be of benefit to society but that some people might vote against compulsory genetic screening in a referendum. **(6 marks)**

Such screening could inform health service planning and identify people at risk. This might lead to early treatment or determine the appropriate dose of medication. People with defective genes could be advised about having children, reducing cost to society. People without genetic defects may benefit from reduced insurance premiums. However, some may feel this is undue intrusion into people's lives by governments. Also, it may be easier to cope if you don't know in advance about a condition. Those with defects may have to pay more for their insurance, or not even be able to get cover. There may be pressure to have abortions.

Drug development

Pharmacogenomics is the science of how an individual responds to a drug because of their genes. Pharnacogenomic tests could be used to:

- identify whether Herceptin will be effective against a person's breast cancer
- determine the correct dose of the blood-thinner warfarin
- find out whether the AIDS drug ABacavir would be effective.

Now try this

Explain what still needs to be done now that the base sequences for all human genes are known. **(2 marks)**

Genetically modified organisms

Drugs can be made by modifying animals, plants and microorganisms to act as drug factories.

GMOs

- GM plants may be useful for producing edible drugs, such as vaccines, that can be stored and transported easily in plant products such as bananas or potatoes.

- Useful genes can be transferred into crop plants using a vector such as *Agrobacterium tumefaciens*, gene guns (pellets coated with DNA) or a virus.

- Restriction enzymes are used to cut DNA at specific sequences and DNA ligase is an enzyme that can be used to stick pieces of DNA together. These make it possible to insert specific DNA sequences into the GM organism. Large numbers of identical GM plants can easily be produced.

- Transgenic animals (animals with a human gene added to them) can be used to produce useful drugs that can be harvested from their milk (or even semen).

- Liposomes and viruses are vectors used to insert genes into animal cells. Drugs produced from transgenic animals include the blood clotting factors used to treat haemophilia.

- Microorganisms, such as bacteria, are the most common target for genetic modification as they are relatively easy targets for gene transfer and can be grown rapidly in large quantities in fermenters. The drugs produced can be extracted and purified. Insulin, to treat type II diabetes, is an example of a drug produced from genetically modified microorganisms.

Introducing genes

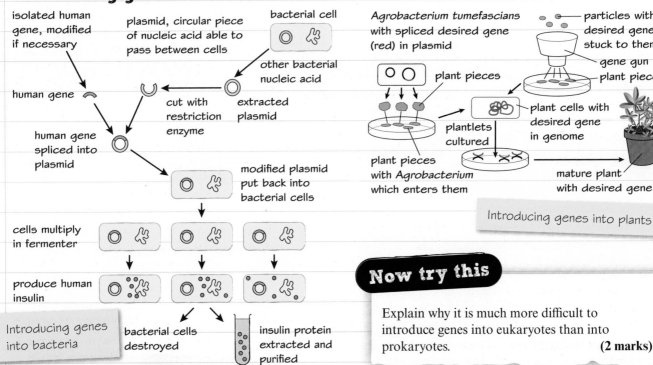

Introducing genes into plants

Introducing genes into bacteria

Now try this

Explain why it is much more difficult to introduce genes into eukaryotes than into prokaryotes. **(2 marks)**

Worked example

Explain some of the concerns over the use of GMOs. **(4 marks)**

Antibiotic-resistant genes are used to identify GM bacteria, which could lead to antibiotic resistance developing in other microbes.

GM crops could become super-weeds that out-compete other plants and may be resistant to herbicides.

They could damage natural food chains, resulting in damage to the environment because they would encourage farmers to use herbicides to kill everything but the crop.

GM crops may not produce fertile seeds. This prevents farmers collecting seed and replanting, so they need to return to the biotechnology company to buy new seeds for each planting. This could make them too expensive for some farmers.

Exam skills

This exam-style question uses knowledge and skills you have already revised. Have a look at page 125 to help you.

Worked example

In an experiment, fifty 25 mm lengths of plant stem were cut and ten were placed in each of five dishes.

A different concentration of IAA was added to each dish. The dishes were left for 24 hours and the mean increase in stem length was recorded.

The results are shown in the table below.

Dish	IAA concentration / mg dm^{-3}	Mean increase in stem length / mm	2 × SD (95% confidence interval)
1	0.00	2.5	1.0
2	0.01	2.1	1.5
3	0.10	5.0	1.1
4	1.00	6.8	4.0
5	10.00	7.8	3.2

(a) From the data given, calculate the likely length of the longest stem piece after 24 hours in 0.10 mg dm^{-3} IAA. **(2 marks)**

The original pieces were 25 mm. In 0.10 mg dm^{-3} solution, the increase was by a maximum of (5 mm + 1.1 mm). So the longest piece would be 25 + 6.1 mm = 31.6 mm.

(b) Plot a graph of the given data on the grid below. Do not include the SD. **(2 marks)**

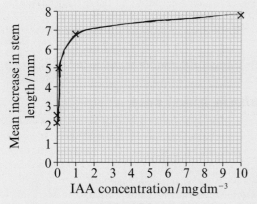

(c) Analyse the data given to explain the effect of IAA concentration on mean stem length. **(5 marks)**

As IAA concentration increases, the increase in mean stem length also increases. However, the rate of increase begins to slow at about 0.05 mg dm^{-3}. Above 1 mg dm^{-3}, the increase is minimal up to 10 mg dm^{-3}.

IAA brings about cell expansion by 'loosening' the cell wall. It causes the acidification of the wall by stimulating the activity of proton pumps that move H$^+$ ions out of the cytoplasm and into the cell wall. The low pH activates proteins called expansins, which disrupt the bonds that hold the cellulose microfibrils together. Water uptake by osmosis, causes the cell to swell and elongate. The loosening and/or uptake of water would eventually reach the maximum possible value.

It is always important that you carefully read the stem of a question. In those which involve figures, some of which may be used in a mathematical calculation, it is even more important that you are clear what is being asked. Here, one very commonly tested skill is being examined: the ability to select the relevant data from all that given. You need to focus on the row for dish 3. Next, you must take into account the phrase 'from the data given'. This tells you the mean increase in length for these stem pieces was 5 mm. The 95% confidene interval means that you can be 95% sure that no length will be greater or less than the mean plus the value, or the mean minus the value. So, the longest piece will be 5 mm plus 1.1 mm longer than the starting length with 95% certainty, addressing the 'likely length' asked for in the question. Also note that column three is increase in length **not** new length.

Maths skills When plotting graphs, accuracy of positioning of the points is crucial. In this case the axes are given, but if they were not you should remember that the independent variable always goes along the x-axis. You also label axes with the variable name and its units if appropriate.

In **analyse to explain** questions you will need to describe the situation in front of you and then apply your own knowledge to make a judgement and reach a conclusion.

Be careful! This is not growth because there would be no increase in dry biomass.

Answers

Topic 1

1. Why is transport needed?

A water molecule has a positively charged hydrogen end and a negatively charged oxygen end. **(2)**

2. Blood vessels

capillary walls are one cell thick whereas veins have many multi celled layers.
Capillaries do not contain no elastic tissue but veins do. **(2)**

3. The heart

The (right) atrium has less muscle than right ventricle. **(1)**
Because it does not need to generate such high blood pressure required. **(1)**
The right atrium pumps blood to right ventricle which is nearby but the right ventricle pumps blood to lungs which are further away. **(1)**

4. The cardiac cycle

These valves separate the atria from the ventricles. They open during atrial systole so that blood can pass through to ventricles. They then close during ventricular systole to prevent blood being forced back into the atria. They open during diastole so that ventricles can start to fill up as atria are filling. **(4)**

5. Clots and atherosclerosis

Damage to endothelial cells causes an inflammatory response. White blood cells and cholesterol accumulate in the inflamed area. A plaque is formed and elasticity of the arteries is lost. **(4)**

6. Risk

High blood cholesterol leads to fatty deposition in artery walls. This leads the lumen of the coronary arteries to narrow.
High blood pressure damages the lining of the arteries and an increased risk of blood clot blocking coronary arteries.
Smoking increases blood pressure and increases the risk of aneurysm. **(5)**

7. Correlation and causation

Experiments in which cigarette smoke, or components of cigarettes which could get into the body, are shown to damage cells and turn them cancerous. Then, we could say '*Smoking greatly increases the chance of developing lung cancer (correlation) because tars and other substances in the smoke damage cells in the lungs (cause)*'. **(3)**

8. Studying the risks to health

(a) 4:3 **(1)**
(b) There is a correlation between being overweight and gender in that a higher percentage of males than females are overweight in every country. For obesity there is no such clear correlation across the countries. **(2)**

9. Energy budgets

$$25 = \frac{100}{x^2}$$

$$x^2 = \frac{100}{25} = 4$$

$$x = \sqrt{4}$$

$$= 2\,m \qquad \textbf{(2)}$$

10. Monosaccharides and disaccharides

Because a molecule of water (H_2O) is lost when the two hexoses are joined with a condensation bond. So formula is $C_{12}H_{24}O_{12} - H_2O = C_{12}H_{22}O_{11}$. **(2)**

11. Carbohydrates – polysaccharides

Structural feature	Sucrose	Amylopectin
Glycosidic bonds present	✓	✓
Side chains present		✓
Contains more than three glucose monomers		✓

12. Lipids

(a) Ester bond **(1)**
(b) It would lower the pH because fatty acids would be released and acids ionise to give hydrogen ions which lower pH. **(2)**

13. Good cholesterol, bad cholesterol

(a) Because people in France eat same amount of fat but drink more alcohol and smoke more cigarettes than people in the UK. **(2)**
(b) Antioxidants in wine might prevent the oxidation of LDL, which is involved in plaque formation, so the French diet may contain a higher percentage of HDLs, there may be differences in other risk factors not studied. **(2)**

14. Reducing the risk of CHD

6 mg Vitamin C decolourised 1 cm^3 of DCPIP. 11.56 cm^3 of pineapple juice decolourised 1 cm^3 of DCPIP so 11.56 cm^3 of pineapple juice contains 6 mg Vit C. So the concentration is $\frac{6}{11.56} = 0.52\,mg\,cm^{-3}$. **(2)**

15. Medical treatments for CVD

Muscle pain and cataracts are both known side effects. **(2)**

16. *Daphnia* heart rate

The biggest problem in ensuring accuracy is that the high heart rate is difficult to count. One way to deal with this is to cool the *Daphnia* and slow the heart rate. Another way would be to film the *Daphnia* and slow down film in order to counter the heart rate more accurately. Another possibility is to use a strobe light to freeze the motion. The rate of strobing equals the heart rate when it appears to be frozen. **(4)**

Topic 2

18. Gas exchange

The walls of the alveoli and blood capillaries are one cell thick. There is a dense network of capillaries around the alveoli and many alveoli provide a large surface area. All of this maximises the rate of diffusion of gases. **(4)**

19. The cell surface membrane

The membrane has a phospholipid bilayer with the hydrophobic tails of the phospholipids in the centre and the hydrophilic heads on the outside, facing water. There are proteins in the membrane, some on either surface, some within the membrane and some crossing right through it. The membrane will have glycoproteins on the outside and cholesterol within it. **(4)**

20. Passive transport

They could have a small molecular size and be soluble in lipids. It would also help if they were recognised by protein receptors. **(3)**

21. Active transport, endocytosis and exocytosis

Insulin is engulfed in a membrane-bound sac called a vesicle. This then moves to the cell membrane where it fuses with the membrane. This is possibly due to the fluid nature of membranes. The insulin is shed to the outside as the vesicle membrane opens. **(3)**

22. Practicals on membrane structure

The suspension of betalain in water should be shaken before measurement.
The colorimeter is zeroed against a blank of distilled water. The liquid is poured into a cuvette and placed in the colorimeter to measure absorbance or transmission. **(3)**

23. The structure of DNA and RNA

If 32% are cytosine then 32% are guanine with which it always pairs. This leaves $100 - 64\%$ for adenine and thymine, that is 36%. Again adenine always pairs with thymine, so $\frac{36}{2}\%$ of each = 18%.
C = 350×0.32 = 112 bases, G = 112 bases
A and T are both 0.18×350 = 63 bases each
Double check 112 + 112 + 63 + 63 = 350 **(3)**

24. Protein synthesis – transcription

AUGAAGGAAACUGCU **(2)**

25. Translation and the genetic code

mRNA lines up on the ribosome and tRNA picks up specific amino acids.
The tRNA undergoes codon–anticodon bonding with the mRNA on the ribosome.
Peptide bonds between adjacent amino acids are formed and the ribosome moves along the mRNA strand to repeat the process. **(4)**

26. Amino acids and polypeptides

D carboxyl group and the amino group **(1)**

27. Folding of proteins

Hydrogen, ionic and disulfide bonds. **(3)**

28. Haemoglobin and collagen

Proteins which consist of more than one polypeptide chain have a quaternary structure. In haemoglobin there are four such groups, two alpha and two beta; they are held together by H bonds. Additional non-polypeptide groups in the molecule are called prosthetic groups. Haemoglobin has four iron-containing haem groups. **(5)**

29. Enzymes

(a) Q water, R active site
(b) substrate
(c) glycosidic
(d) hydrolysis **(4)**

30. Activation energy and catalysts

A biological catalyst is an enzyme which reduces activation energy of a reaction.
Activation energy is the energy needed for a reaction to occur by causing bonds to break.
The active site is a 3D cleft on the surface of an enzyme. It binds to a protein or other substance during a reaction. **(5)**

31. Reaction rates

There should be enough substrate molecules to saturate the enzyme so that the substrate is not a limiting factor.
As the reaction proceeds, the substrate concentration decreases as substrate gets used up by enzyme.
So, to make sure it is the same in each replicate, the initial rate is measured. **(3)**

32. Initial rates of reaction

Temperature can be controlled in a water bath with a thermostat. pH can be controlled with the use of buffer solutions. **(4)**

34. DNA replication

The adenine and thymine will separate first and then onwards down the other base pairs.
As this 'unzipping' proceeds, individual nucleotides with the relevant base will pair on both strands from the top. So, a T will pair with the A at top left and an A with the T at top right. Next will be a G joining to the C on the left and a C with the G right and so on until two new identical stands are made.
The process will be catalysed by DNA helicase (the unzipping) and then DNA polymerase and DNA ligase to join the nucleotides. **(5)**

35. Evidence for DNA replication

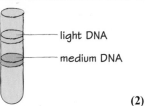

(2)

36. Mutation

Despite the fact that there is a range of problems listed, the final outcome of all is the same. Chloride ions will not be able to move as normal. Therefore sticky mucus will form and this could lead to respiratory, digestive and reproductive problems. **(4)**

37. Classical genetics

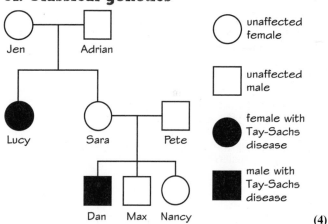

(4)

38. Cystic fibrosis symptoms

Overall there is an increase in *P. aeruginosa* (P) and a decrease in *S. aureus* (S).
Between 0 and 7 years S is greater than P, but after 7 years P is greater than S.
S starts to decrease at year 15, but P decreases at 35 years.
The maximum P is greater than maximum S. **(5)**

39. Genetic screening

Amniocentesis is usually associated with a 1% risk of causing a miscarriage.
It cannot be carried out until about 17 weeks or so into the pregnancy.
CVS can be carried out earlier in pregnancy, between 8 and 12 weeks. It is associated with a slightly higher risk (between 1 and 2%) of inducing a miscarriage than amniocentesis.
Both tests will present the prospective parents with a result that may mean they wish to consider an abortion. In many cases a couple may have thought about this prior to having the test. If so, and their decision is that the woman would have an abortion if the test is positive for a particular condition, then CVS is preferable because an abortion is easier for the woman the earlier it is done. **(6)**

Topic 3

42. Prokaryotes

B and C

C is nucleoid

B is 70S ribosome **(3)**

43. Eukaryotes

Two membranes	One membrane	No membranes	
nucleus mitochondria	lysosomes Rough and smooth ER Golgi apparatus (or Golgi body)	ribosomes centrioles nucleolus	**(3)**

44. Electron micrographs

The packaging and modification of proteins. **(2)**

45. Protein folding, modification and packaging

Proteins are produced by ribosomes on rough ER. The proteins are folded within rough ER.

Vesicles from rough ER fuse with Golgi apparatus. Here the protein is Golgi modified, for example glycoprotein. Carbohydrate is added. The Golgi produces secretory vesicles for secretion of extracellular enzymes. **(5)**

46. Sperm and eggs

The sperm cell has flagellum, the egg does not. The sperm cell has fewer mitochondria than the egg.

The sperm cell has acrosome not found in the egg. **(3)**

47. Genes and chromosomes

They have 8 chromosomes in a body cell, so 4 in a gamete, so 4 different chromosomes.

Genes in a group are on the same chromosome, so they have 4 groups of linked genes. **(2)**

48. Meiosis

The two processes are independent assortment and crossing over. Independent assortment gives rise to different combinations of paternal and maternal chromosomes.

Crossing over involves swapping of sections of chromatids. **(3)**

49. The cell cycle

16 (after 1 division it is 2, then 4, then 8, then 16). **(1)**

50. Mitosis

Ethanoic orcein – to stain the chromosomes

Concentrated acid – to separate the individual cells **(2)**

52. Stem cells and cell specialisation

Totipotent cells can differentiate to become any cell.
Pluripotent cannot differentiate to become all cells in the body.
Only totipotent cells can give rise to other totipotent cells.
Totipotent cells can give rise to an entire human being, pluripotent cells cannot. **(2)**

53. Gene expression

A stimulus such as a chemical causes some genes to become active in the bone marrow stem cell.

Only the active genes are transcribed because mRNA is made only at active genes.

Proteins are made which determine cell structure and function, permanently modifying the cell. **(4)**

54. Nature and nurture

Genetic programming (nature) is likely to be responsible for innate reflexes, because they are not influenced by the environment (nurture). **(2)**

55. Continuous variation

The mean and the median are both 69 cm. **(2)**

56. Epigenetics

RNA polymerase is an enzyme with an active site that is specific to normal DNA bases.

Methylation will change the shape of cytosine, preventing normal action of the enzyme. **(2)**

Topic 4

57. Biodiversity

D for wood = 1.29

D for wheat = 1.04

So data support the conclusion BUT, only for wheat field, not all farming, and only insects not all organisms. Only for woodland not other natural habitat. **(4)**

58. Adaptation to niches

Selection pressure from the hot and dry habitat acted on the frogs. Those with a chance mutation had an advantageous allele for waxy secretions. Those individuals with this advantageous allele would survive and breed. The idea of advantageous allele is passed on to future generations. This means the frequency of advantageous alleles in the population increases. **(6)**

59. Evolution and speciation

Speciation is the formation of new species from existing species. It arises due to reproductive isolation and the new species are no longer able to produce fertile offspring with existing species. **(2)**

60. The classification of living things

They have different-sized ribosomes. Bacteria lack organelles, whereas eukarya lack a slime capsule. **(3)**

61. The validation of scientific ideas

It conflicts with religion and creationism.

It suggests the world is older than given in religious texts.

Some people believe that the evidence for evolution is unsatisfactory and the idea that humans are evolved from apes is considered particularly offensive. **(3)**

62. Plant cells

	Plant (eukaryotic) cell	Animal (eukaryotic) cell	Bacterial (prokaryotic) cell	
Cell wall	✓	✗	✓	
Chloroplasts	✓	✗	✗	
Nuclear membrane	✓	✓	✗	
Cell surface membrane	✓	✓	✓	
Ribosomes	✓	✓	✓	
Centrioles	✗	✓	✗	**(5)**

63. Cellulose and cell walls

Cellulose is flexible and inelastic. A parallel arrangement of several layers of fibres are held together by H bonds in a criss-cross pattern. **(2)**

64. Transport and support

The tissue which transports sugars is the phloem. Phloem is situated towards the outside of the tree bark and therefore ringing will remove it. The sugars which are needed for the production of energy in respiration and as building blocks for new substances will not get to the places that need them. **(5)**

65. Looking at plant fibres

Fibres of the same length and width are taken.
During the experiment temperature and humidity are kept constant.
The masses were added to the fibres in the same way each time.
Masses are added until the fibre breaks and the mass that causes this is noted.
Cushions are laid to catch the falling masses. **(5)**

66. Water and minerals

Magnesium and nitrate are used for production of chlorophyll which is used for photosynthesis. An increase in photosynthesis leads to increased yield.
Nitrate is used in the production of protein and more protein increases growth and thus yield. **(4)**

67. Developing drugs from plants

Contemporary is safer because pure drug is used rather than extract.
It is more valid because a placebo is used as a comparison and also involves double-blind trials.
It is more reliable because more people are tested and the results are analysed statistically. **(4)**

68. Investigating antimicrobial properties of plants

(a) $5.54^2 / 5 = 6.14$ and $6.19^2 / 5 = 7.66$
$\sqrt{(6.14 + 7.66)} = 3.71$
$(21.7 - 22.2) / 3.71 = t = -0.13$ **(3)**

(b) The t-test assesses the significance of the difference between the means of the two treatments. Chi squared not appropriate because there are no expected values. Correlation coefficient not appropriate because the independent variable is discontinuous / not continuous. **(3)**

69. Conservation: zoos

18 individuals is a small population with a small gene pool, captive breeding will increase the population.
Studbooks kept of the breeding programme are used by zoos to select mates. This may mean inter-zoo exchange of animals for breeding. In some cases IVF or AI can be used. **(6)**

70. Conservation: seed banks

They would be taken from different plants to provide genetic variation. They would then be X-rayed to check for viability. (3)

Topic 5

72. Ecosystem ecology

Different mouth shapes are likely to be related to different foods. This will mean that interspecific competition will not occur so each species can exist in its own niche. **(2)**

73. Distribution and abundance

Light intensity was measured using a light meter above vegetation and at the base of it. Several readings were taken at each position and a mean and SD calculated. **(3)**

75. Succession

It includes both animals and plants which interact to give a balanced equilibrium of species. It contains dominant species and is stable if there is no change to the environment. **(3)**

76. Productivity in an ecosystem

The rate of chemical reactions increases with temperature, due to an increase in kinetic energy of enzyme and substrate molecules. This leads to an increase in reaction rate because of more enzyme substrate interactions. **(2)**

77. Energy flow

(a) NPP = 2381.0
Input = 3 144 000
Efficiency = 2381 / 3 144 000 × 100% = 0.08% **(3)**

(b) Because the estimate of NPP ignores energy stored in roots, and the material eaten by herbivores or lost in leaf fall etc., over the 20 years. **(2)**

78. Photosythesis: an overview

S hydrolysis
T phosphorylation
W phosphate **(3)**

79. The light-dependent reactions

A ATP
B NADPH
C water **(3)**

80. The light-independent reactions

D RuBISCO **(1)**

82. Chloroplast

C release of H^+ from water **(1)**

83. Climate change

1960 figure is 310 ppm and in 2010 390 ppm, the difference is 80, so percentage change is 80 / 310 × 100 = 25.7%, which is an increase of 25.8% **(2)**
Answers may vary according to the figure used for 1960.

84. Anthropogenic climate change

Global warming means just an increase in environmental temperature.
The greenhouse effect is something which causes global warming. It is a theory to explain changes in global temperature, which suggests that heat is being trapped in the atmosphere by carbon dioxide and other greenhouse gases. **(3)**

85. The impact of climate change

Fewer prey are eaten by tuataras, so prey increase in number. This means other carnivores may increase because there is less competition for food from tuataras. A predator of tuatara might decrease. **(2)**

86. The effect of temperature on living things

carnation sedge. **(1)**

87. Decisions on climate change

Land has to be cleared to grow plants for biofuels and the plants growing there would have been using carbon dioxide already. The land could have been used for food production, so now less food is produced. **(4)**

88. Evolution by natural selection

White cats will be selected against as their deafness will disadvantage them. The white allele will not be passed on and thus reduce in frequency so the equilibrium will not be seen. **(3)**

89. Speciation

Allopatric because they are in different places (other mosquitoes are not in the underground). **(2)**

Topic 6

90. Death, decay and decomposition

There is a delay of 4 hours before rigor starts due to ATP still being in the body. Then rigor increases from 4 until 14 hours due to all ATP having been used. Finally there is a decrease in rigor after 14 hours due to the muscle tissue breaking down. **(3)**

91. DNA profiling

Polymerase used in PCR to amplify DNA in small samples.
Restriction endonuclease use to create DNA fragments. **(4)**

93. Bacteria and viruses

Penicillin, because viruses do not have a cell wall.
Tetracycline, because viruses do not make proteins. **(2)**

94. Pathogen entry and non-specific immunity

Interferon is involved in viral infections; lysozyme affects bacteria.
Interferon is produced by infected cells; lysozyme is present in secretions.
Interferon inhibits the replication of viruses; lysozyme kills bacteria.
But both protect against infection due to the entry of a pathogen. **(3)**

95. Specific immune response – humoral

B cells target particles and soluble antigen, T cells target cellular material.
B cell effector is a plasma cell, T cells is T killer cell.
Plasma cells secrete antibody to destroy antigen, T killer causes cell lysis. **(3)**

96. Specific immune response – cell-mediated

Cytokines from T helper cells are involved in activating B cells and thus antibody production by plasma cells.
They are involved in the cell-mediated response activating T killer cells which destroy infected host cells. **(3)**

97. Post-transcriptional changes to mRNA

U G C C	A U G	U U C	G G C	G G A	U A C	U A G	C
	Met	Phe	Gly	Gly	Tyr	Stop	
	+ start codon						

(3)

98. Types of immunity

Antibodies in breast milk provide passive immunity but last a short time.
The vaccine will produce secondary response and memory cells which remain in circulation for many years. **(4)**

99. Antibiotics

A nutrient agar plate is inoculated with a bacterium. The antibiotic is applied using sterile paper discs. The plate is incubated at a temperature below 30 °C to prevent growth of human pathogens. Aseptic technique such as flaming is used to prevent contamination. Antibiotic effectiveness can be measured after 36–48 hours by finding the diameter of the inhibition zone. **(5)**

100. Evolutionary race

Changes in humans, such as the immune system preventing HIV growth, acts as a selection pressure on HIV.
Only those particles adapted will replicate. These become more common. **(4)**

Topic 7

102. Skeleton and joints

A   The tendon attached to the flexor muscle.
B   The tendon attached to the extensor muscle.
C   Biceps muscle.
D   The tendon attached to the extensor muscle, the triceps. **(4)**

103. Muscles

The sarcoplasmic reticulum contains calcium ions which bind to troponin. This causes tropomyosin to move, exposing binding sites for myosin. ATP removes the calcium ions, changes the shape of myosin and breaks cross bridges. **(5)**

104. The main stages of aerobic respiration

NAD accepts hydrogen when glucose is oxidised in glycolysis. NAD accepts hydrogen when compounds are oxidised in Krebs' cycle. FAD is reduced during one of the reactions in Krebs' cycle. NAD and FAD are oxidised when they pass hydrogen ions and electrons to molecules of the ETC. The hydrogen ions and electrons provide the energy to produce ATP during reactions of the ETC and by chemiosmosis. **(5)**

105. Glycolysis

(a) Phosphorylation is the addition of phosphate to a molecule. Phosphorylation of glucose of is an early stage of glycolysis. It converts glucose into a more reactive molecule so that the process of glycolysis can proceed. **(3)**

(b) Glycolysis involves oxidation of glucose by removing hydrogen atoms. NAD is a coenzyme needed to accept the hydrogen atoms in order that the reaction can occur. Reduced NAD can enter the mitochondrion and pass hydrogen atoms to the ETC and this results in ATP production. **(3)**

106. Link reaction and Krebs cycle

It is acetyl coenzyme A that combines with a 4-carbon compound produced by reactions of Krebs' cycle to form a 6-carbon compound that is the starting material for other reactions of Krebs' cycle. Without coenzyme A, this reaction would not be possible and the reactions of Krebs' cycle could not continue. Pyruvate produced in the link reaction would build up as it would not be used. Carbon dioxide production would stop and there would be no reduction of NAD and FAD in Krebs' cycle which is necessary for the ETC to function. ATP production would fall significantly. **(3)**

107. Oxidative phosphorylation

(a) Glycolysis occurs in the cytoplasm, and mitochondria lack the enzymes needed to convert glucose to pyruvate. Enzymes of the link reaction use pyruvate as a substrate in the mitochondrial matrix. **(2)**

(b) Proteins of the ETC are present in the mitochondria and will accept hydrogen from reduced NAD, passing H$^+$ into the intermembrane space as usual. This oxidises the reduced NAD. H$^+$ will be able to follow a diffusion gradient into the matrix without passing through ATP synthase which means that ATP will not be produced by the flow of ions. **(4)**

108. Anaerobic respiration

(a) Anaerobic respiration produces ATP without the need for oxygen. Glycolysis is inefficient but will produce ATP as glucose is converted to pyruvate. NAD is reduced in the process but is oxidised by reducing pyruvate to lactate. This requires large quantities of glucose which is available from stores of glycogen in the muscle fibres. **(6)**

(b) The build up of lactate will cause a fall in blood pH which affects enzymes and muscle contraction resulting in fatigue. **(2)**

109. The rate of respiration

Rates are given by distance in 15 mins so for 12 mm it is 12/15 = 0.8 mm min^{-1}, other values give 0.73 and 0.87 and then 0.80 again, so mean is 3.2/4 = 0.80 mm min^{-1}.
SD: x – mean for 12 is 0.8 − 0.8 = 0 for 11 it is 0.73 − 0.8 = −0.07, for 13 it is 0.87 − 0.8 = 0.07 and then 0 again; square each of these to give 0 + 0.0044 + 0.0049 + 0 = 0.0078; divide by n − 1 (=3) gives 0.0033; square root of 0.0033 is 0.057 which is the SD. **(4)**

110. Control of the heartbeat

There is a delay at the AV node. The impulse is then carried across the bundle of His and along the Purkyne fibres to the base.
The impulse travels from base up through and stimulates contraction of cardiac muscle. **(3)**

111. Cardiac output and ventilation rate

Heart rate and stroke volume increase. Also, SAN activity increases as well as the AVN time delay decreases.
More blood returns to the heart, which causes heart muscle to stretch which leads to the ventricles contracting with greater force. **(4)**

112. Spirometry

The increase in heart rate seems to be the main cause of the increased oxygen uptake as there is only a small difference in ventilation rate, which actually goes down at the higher heart rate. **(3)**

113. Fast and slow twitch muscle

B more myoglobin than fast twitch fibres **(1)**

114. Homeostasis

Enzymes are sensitive to pH and temperature.
Intracelleuar enzymes are very intolerant of change.
pH can affect the 3D shape of enzymes as it affects hydrogen bonds holding it in place.
The charge distribution on active sites will also be affected causing less effective binding with substrate through induced fit.
Temperature affects enzyme activity as it affects energy in particles changing the rate of enzyme substrate collisions.
At high temperatures the enzyme may denature. **(4)**

115. Thermoregulation

This is an example of homeostasis using a negative feedback mechanism. Changes to the core temperature are detected by thermoreceptors in the hypothalamus which send nerve impulses to arterioles in the skin. This causes vasodilation resulting in increased blood flow to the skin. **(4)**

116. Exercise

Pathogens are encountered through runners meeting from other areas. The weakened immunity due to a fall in natural killer cells makes them susceptible to airborne infection. **(3)**

117. Sports participation and doping

Relativists say that drugs could be used under some circumstance. (for example, they could be used for medication) and that drugs in the body can be difficult to legislate for. **(3)**

Topic 8

119. Mammalian nervous system

Antagonistic means working in opposition to each other.
Changes made by the sympathetic system can be revesed by the parasympathetic system, for example.
They allow changes to be made, for example in regulating the rate of heart beats according to conditions. **(3)**

120. Stimulus and response

Receptor: {Rod / cone cells / retina / photoreceptors} **(1)**
Effector: Iris muscle / radial and / or circular muscle **(1)**

121. The resting potential

The phospholipid bilayer prevents movements of sodium and potassium ions as they are polar.
The sodium–potassium pumps are proteins in the membrane which maintain high concentration gradients for sodium and potassium ions.
Potassium channels allow potassium ions to leave the cell following the diffusion gradient. **(3)**

122. Action potential

Potassium ions leave the cell following diffusion gradient, and are attracted by negative charge. **(4)**

123. Propagation of an action potential

refractory period is when the axon membrane cannot depolarise threshold potential caused by influx of sodium ions will only depolarise membrane towards direction of conduction and will not return **(4)**

124. Synapses

EPSPs cause opening of sodium ion channels to cause reduction in potential difference of the postsynaptic membrane.
IPSPs cause hyperpolarisation of the postsynaptic membrane.
EPSPs make it more likely that the threshold potential will be reached causing a postsynaptic action potential, IPSPs make reaching the threshold less likely. **(3)**

125. Vision

ATP (supplies energy) for active transport/reference to sodium–potassium pump.
This pumps sodium ions out of inner segment/maintains (more) negative charge inside the membrane. **(2)**

126. Plant responses

Both chemicals are produced in cells;
both move away from site of production;
effect may be distant from production site / eq;
long-term / permanent effect / example quoted / eq;
involved in gene activation /eq. **(3)**

127. Nervous and hormonal control

For example,
Heart rate adrenaline acts directly on heart to speed it up nerves (both sympathetic and parasympathetic) innervate the heart to speed it up and slow it down. **(3)**

128. The human brain

CT can only identify larger structures; MRI can identify smaller structures.
MRI uses magnetic field; CT uses X-rays.
both give 3D images; at one point in time. **(4)**

129. Critical window for development

Animal experiments help to test medicines and treatments. On a utilitarian philosophy the expected benefits are greater than expected harms. Also, these experiments reduce the chances of harm when testing on people. **(3)**

130. Learning

The snail will ignore unimportant stimuli.
More receptive to important stimuli.
Less time wasted with tentacles withdrawn/covered.
More time to search for food. **(4)**

131. Brain development

Identical twins are genetically identical/eq
Height mainly due to {genes/nature}/not affected much by environment
Body mass and intelligence are mainly due to {environment/nurture/eq}
Reference to figure(s) to back up argument/valid comparison
e.g. less effect of nurture of those reared together/converse
Reference to body mass e.g. diet, exercise
Reference to intelligence e.g. schooling, parental encouragement
Reference to height being {polygenic/multifactorial} **(4)**

132. Brain chemicals

Dopamine agonists mimic dopamine. They bind to dopamine receptors at synapses.
They trigger action potentials. **(3)**

133. HGP – the Human Genome Project

To identify all the proteins coded for by these genes, and discover their functions. **(2)**

134. Genetically modified organisms

The genetic material of eukaryotes is inside a nucleus, so the delivery method has to get the introduced gene through the nuclear membrane. **(2)**

Published by Pearson Education Limited, 80 Strand, London, WC2R 0RL.

www.pearsonschoolsandfecolleges.co.uk

Copies of official specifications for all Pearson qualifications may be found on the website: qualifications.pearson.com

Text © Pearson Education Limited 2016
Typeset by Kamae Design
Produced by Out of House Publishing
Illustrated by Tech-Set Ltd, Gateshead
Cover illustration by Miriam Sturdee

The right of Gary Skinner to be identified as author of this work has been asserted by him in accordance with the Copyright, Designs and Patents Act 1988.

First published 2016

19
10 9 8 7 6 5 4

British Library Cataloguing in Publication Data
A catalogue record for this book is available from the British Library

ISBN 978 1 447 99271 4

Printed in Slovakia by Neografia

Acknowledgements
The publisher would like to thank Edexcel for permission to reproduce extracts from the following exam papers: Jan 2013; May 2014; June 2009; May 2010; Jan 2009; May 2014; Jan 2012; Jan 2010; Jan 2009; June 2005; June 2007; June 2006; Jan 2014; Jan 2014; May 2011; June 2013; Jan 2011; June 2014; June 2007; June 2014; Jan 2004; Jan 2010; June 2005; June 2011; June 2009; June 2005; June 2006; June 2008; Jan 2015.

The author and publisher would like to thank the following individuals and organisations for permission to reproduce photographs:

(Key: b-bottom; c-centre; l-left; r-right; t-top)

Alamy Images: Rebecca / Stockimo 119br, Zoonar GmbH 119bl
Getty Images: Ed Reschke 50
Science Photo Library Ltd: 44b, ASTRID & HANNS-FRIEDER MICHLER 62r, BILL LONGCORE 44/1, BIOPHOTO ASSOCIATES 44/6, 62l, CNRI 2, DON FAWCETT 44/2, DON W. FAWCETT 44/5, DR DON FAWCETT 44/4, DR. GLADDEN WILLIS, VISUALS UNLIMITED 112t, Microscape 44/3, POWER AND SYRED 18, STEVE GSCHMEISSNER 51, STEVE GSCHMEISSNER 51, THOMAS DEERINCK, NCMIR 102
University Hospital Southampton NHS Foundation Trust: 127l, 127br

We are grateful to the following for permission to reproduce copyright material:

Figures
Figure on page 71 from Fig 2: Competitive Relationships Between Certain Species of Fresh-Water Triclads, Journal of Ecology, Vol 20, pp200-208 (R. S. A. Beauchamp and P. Ullyott 1932), Reproduced with permission of John Wiley & Sons, Inc.; Figure on page 82 adapted from Issues in Environmental Science and Technology, Volume 38 (Hester, R and R Harrison 2014), Cambridge, GBR:, Reproduced by permission of The Royal Society of Chemistry; Figure on page 82 adapted from data provided by the Met Office, http://www.metoffice.gov.uk/hadobs/hadcet/data/download.html; Figure on page 100 from Fig 3b: Discrimination of Anemonefish Species by PCR-RFLP Analysis of Mitochondrial Gene Fragments, EnvironmentAsia, Vol 1, No 1, pp51-54 (Chuta Boonphakdee and Pichan Sawangwong), Thai Society of Higher Education Institute on Environment; Extract 4 on page 60 adapted from Ernst Mayr and the modern concept of species, Colloquium, vol.102, suppl.1 (Kevin de Queiroz), "Copyright (2005) National Academy of Sciences, U.S.A."

All other images © Pearson Education